T0305854

Fundamentals of Evapotranspiration

Fundamentals of Evapotranspiration aims to determine simple methods to evaluate evapotranspiration and to examine the evolution of these methods over time. It compares and contrasts best practices and discusses the opportunities for harmonization among various methods. Further, the book discusses optimal calibration of these methods in a local context, depending on particular climates and scenarios. The book serves as a practical resource for students and professionals working in agriculture, irrigation and water engineering and will aid in evaluating the methods and equations for the most efficient means of evapotranspiration.

- The authors examine the methods for evaluating evapotranspiration considering evaporation from water surfaces, soil and vegetation.
- The authors address issues according to various regions, climates and soil types and apply the optimal solution for each situation.

Fundamentals of Evapotranspiration

Alessia Corami, Saeid Eslamian,
and Faezeh Eslamian

CRC Press
Taylor & Francis Group
Boca Raton London New York

CRC Press is an imprint of the
Taylor & Francis Group, an **informa** business

Designed cover image: Shutterstock

First edition published 2025
by CRC Press
2385 NW Executive Center Drive, Suite 320, Boca Raton FL 33431

and by CRC Press
4 Park Square, Milton Park, Abingdon, Oxon, OX14 4RN

CRC Press is an imprint of Taylor & Francis Group, LLC

© 2025 Alessia Corami, Saeid Eslamian, and Faezeh Eslamian

Library of Congress Cataloging-in-Publication Data
Names: Corami, Alessia, author. | Eslamian, Saeid, author. | Eslamian,
    Faezeh A. author.
Title: Fundamentals of evapotranspiration / Alessia Corami, Saeid
    Eslamian and Faezeh Eslamian.
Description: First edition. | Boca Raton, FL : CRC Press, 2025. |
    Includes bibliographical references and index.
Identifiers: LCCN 2024020451 | ISBN 9781032737034 (hbk) |
    ISBN 9781032740126 (pbk) | ISBN 9781003467229 (ebk)
Subjects: LCSH: Evapotranspiration.
Classification: LCC S600.7.E93 C67 2025 | DDC 551.57/2—dc23/
    eng/20240731
LC record available at https://lccn.loc.gov/2024020451

ISBN: 978-1-032-73703-4 (hbk)
ISBN: 978-1-032-74012-6 (pbk)
ISBN: 978-1-003-46722-9 (ebk)

DOI: 10.1201/9781003467229

Typeset in Times
by Apex CoVantage, LLC

Contents

About the Authors

Alessia Corami completed her degree in earth science at the University of Rome La Sapienza with her dissertation "Characterization of mortars and plasters from the archaeological site of Elaiussa Sebaste (Turkey) and from Quintili's Villa on the ancient Appian way in Rome in 1998." Later, she won a fellowship regarding isotopic chemistry at Geokarst srl AREA Science Park, analyzing organic and inorganic samples. In 2001, she won a fellowship at the Georesources and Geoscience Institute of Italian CNR for "Investigation on ancient emeralds to determine the original mine." In 2002, she won a Ph.D. fellowship at University of Rome La Sapienza for "Phosphate-induced heavy metals immobilization in aqueous solutions and soils." During her Ph.D. work, she spent seven months in the United States at Miami University (Ohio) working with Professor John Rakovan, attending the GLG 699.E course "Scanning Probe Microscopy: Theory and Application," and analyzing samples with AFM under Professor Rakovan's supervision. Later, she spent a period at University of California, Merced, working with Professor Samuel J. Traina and analyzing samples at high temperature and high pressure. In 2005, 2006 and 2008, she went to ESRF in Grenoble, working with Dr. D'Acapito and analyzing her Ph.D. samples. From 2007, she was a professor at Guglielmo Marconi University for five years, teaching two courses, Environmental Geology and Geomorphology. In 2014, she was a visiting professor at Karl Eberhard University of Tübingen (Germany) teaching (Bio)-remediation Technologies and Climate Change. During her career, she published articles about archaeometry, geochemistry and remediation techniques. She is a thesis supervisor and reviewer, and she is on the international editorial boards for some scientific journals as well as a member of many scientific organizations. After two years as visiting assistant professor at University of Louisiana, Lafayette she joined le Cerege at European Centre Research and Teaching in Geosciences de L'environment Technopôle de l'Arbois-Méditerranée.

Saeid Eslamian received his Ph.D. in civil and environmental engineering from University of New South Wales, Australia, in 1998. Saeid was a visiting professor at Princeton University and ETH Zurich in 2005 and 2008, respectively. He has contributed to more than 1,000 publications in journals, conferences and books. Eslamian was honored as a 2-Percent Top Researcher by Stanford University for several years. Currently, he is a full professor of Hydrology and Water Resources and the Director of the Excellence Center in Risk Management and Natural Hazards. Isfahan University of Technology. His scientific interests are floods, droughts, water reuse, climate change adaptation, sustainability and resilience.

Faezeh Eslamian is a Ph.D. holder of bioresource engineering from McGill University. Her research focuses on the development of a novel lime-based product to mitigate phosphorus loss from agricultural fields. Faezeh completed her bachelor's and master's degrees in civil and environmental engineering from Isfahan

University of Technology, Iran, where she evaluated natural and low-cost absorbents for the removal of pollutants such as textile dyes and heavy metals. Furthermore, she has conducted research on worldwide water quality standards and wastewater reuse guidelines. Faezeh is an experienced multidisciplinary researcher with research interests in soil and water quality, environmental remediation, water reuse and drought management.

Preface

The estimation of evapotranspiration parameters is fundamental for effective water balance and water resource management. Proper water management reduces the loss of water and allows for optimizing agricultural water applications. Because of increasing populations, water use is increasing greatly, and a large percentage of potable water is being used for agriculture, therefore, water use for agriculture is of a great concern. The two processes in evapotranspiration, soil evaporation and crop transpiration, are very difficult to estimate separately to arrive at a correct equation that considers the climate and the kind of crop. Hence, the subject warrants thorough study.

The aim of this book is to review the different equations for estimating evapotranspiration, highlighting the advantages and disadvantages for each equation, the importance of climate and most important, the relevant parameters for the best-suited equation for the specific climate in a given region. The book aims to give readers a deep understanding of the relevance of crop water and how evapotranspiration equations are fundamental tools for saving water; the authors evaluate these equations starting with FAO56PM. The authors also investigate evapotranspiration and the consequences of global climate change.

This book is fundamental for professionals working in water and irrigation engineering and agronomists in determining which evapotranspiration equation will save most the water. The book is also a tool for undergraduate students, to give them the evapotranspiration equations in a convenient format and help them understand the differences, the advantages and the evolution of each equation. It is fundamental for hydrology and irrigation engineering courses. It is an easy manual for both students and professionals.

The key benefits of our book are that it is 1) a helpful book for beginners; 2) a comprehensive book for relatively accurate estimation of evapotranspiration depth; 3) a mini-book for practitioners in farm water management; 4) an updated book that uses learning machine models; and 5) a user-friendly book for evapotranspiration application in various climates.

Alessia Corami
Cerege Centre de recherche et d'enseignement des géosciences de l'environnement/AixMarseille Université France

Saeid Eslamian
Isfahan University of Technology, Iran

Faezeh Eslamian
McGill University, Canada

1 Introduction

Water cycles worldwide have been altered because of global warming, increasing droughts and flooding have changed water resource distributions. In the last 50 years, potential evapotranspiration (ET_0) has decreased according to the pan evaporation paradox (Peterson et al., 1995; Liu and Zhang, 2011; Ma et al., 2012; Li et al., 2022). Generally, ET_0 is affected by climate, topography, and vegetation; so that not every equation is appropriate for every region or type of vegetation.

Evapotranspiration is considered the main component of the hydrologic cycle concerning the water used for agriculture. Agricultural water use planning and management are becoming fundamental due to increasing populations, particularly in arid and semiarid regions (Pereira et al., 1999; Mpusia, 2006; Kamali et al., 2015; Amiri et al., 2019). ET and irrigation management are also becoming important for avoiding the contamination of groundwater (Amiri et al., 2019). According to Lingling et al., (2013) – ET is a fundamental element of energy balance. Estimating ET on hydrological models requires considering the energy influence according to time and space to determine the accurate amounts of water necessary (Xu et al., 2000, 2003).

ET is considered to mostly account the water balance in dry areas (Wilcox et al., 2003). In agriculture, water management practices such as irrigation systems allows for forecasting crop yields (Allen et al., 1998). Lysimeters allow for estimating ET in a crop area, but mathematical models are economic alternatives; however, their results are empirical and not always precise, and they might be worst in arid and semiarid regions, better results come from physical models. Empirical models refer to statistical functions with meteorological variables and ET values (Thornthwaite, 1948; Blaney and Criddle, 1950; Jensen and Haise, 1963; Hargreaves and Samani, 1985; Landeras et al., 2008), whereas physical models use two fundamental factors: energy for latent heat evaporation and water vapor flux.

Evapotranspiration equations are algorithms based on weather data, influenced also by surface energy and vegetation aerodynamic characteristics, resulting in a nonlinear phenomenon. These two features are deeply variable and not well defined in case of forest, desert and riparian systems. Both, models and calibrations, can contain errors due to experimental designs, measurement equipment, vegetation management, parameterization and data interpretations (Allen et al., 2011). In ET models, the procedures and equipment for measuring weather data are fundamental for parameterizing hydrologic and weather data and better operating water management in the face of climate change. Lysimeters are effective for measuring ET, but their use is a very long procedure, and a strictly accurate design is necessary for obtaining reliable results (Allen et al., 1998, 2011). FAO56PM is considered the most accurate equation for estimating evapotranspiration. Unfortunately, it requires a great deal of meteorological data such as solar radiation, wind speed and relative humidity that in many cases cannot be easily measured, in both developed and undeveloped countries.

DOI: 10.1201/9781003467229-1

Crop evaporation should not exceed free surface water evaporation; to prevent this, Allen et al., (2011) suggested evaluating ET measurements and possibly rejecting the data. Droogers and Allen, (2002) stated that FAO56PM is better than the other methods because it is based on a physical approach and can be used everywhere, and most important, it is well documented. ET_0 maps are based on the weather data from meteorological stations, but unfortunately the accuracy depends on the climate. ET_0 prediction is fundamental for irrigation at a regional scale, but its accuracy depends on the spatial variability (Zhao et al., 2005).

In arid and semiarid regions, the use of greenhouses is suggested for avoiding ET; the use of greenhouses changes the radiation balance in the environment and prevents moisture loss (Fernandes et al., 2003; Amiri et al., 2019). Because the accuracy of FAO56PM is sometimes hampered by scarce meteorological data, simpler ET equations with fewer variables have been tested. Data from the empirical models needs to be calibrated from FAO56PM, and generally, these models show reliable results.

Partitioning ET into evaporation from the soil and transpiration from the plants is fundamental, but more important is to assess the biomass amount and the knowledge of the areas where water is scarce (Kool et al., 2014). There are many methods evaluating ET, starting from considering the evaporation from single surfaces such as water, bare soil and vegetation; these methods statistically estimate the amount of water but do not consider the hydrological cycle (Lingling et al., 2013).

The aim of this chapter is to determine simple methods for estimating ET and to evaluate these methods. Pereira et al., (1999) stated that the theoretical knowledge must be validated through field measurements. It is necessary to synthesize the methods of estimating ET because it is fundamental in determining the availability of water. However, Rácz et al., (2013) highlighted that harmonizing these methods can be challenging, suggesting the opportunity to compare and calibrate these methods in local context.

1.1 EVAPORATION DEFINITION

Evaporation is a process whereby water (liquid) become vapor, decreasing as it evaporates from surfaces such as lakes, rivers, pavements and soil. This process needs energy to change the molecular state from liquid to vapor, energy partially from solar radiation and partially from the air temperature. The driving force is the difference between the water vapor pressure at the water surface and the air pressure. As evaporation continues, the air becomes saturated, and the process slows down (Allen et al., 1998).

1.2 TRANSPIRATION DEFINITION

Transpiration is the vaporization of water (liquid) from plant tissue to the atmosphere. Generally, crops lose water through stomata (small openings on the leaf), water is taken up from the roots, and mostly it is lost because of transpiration. The energy is from the vapor pressure and wind; therefore, air temperature, air humidity and winds are important in measuring transpiration (Allen et al., 1998).

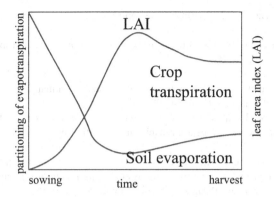

FIGURE 1.1 The partitioning of evapotranspiration into evaporation and transpiration over the growing period for annual field crops.

Source: Modified from Allen et al. (1998).

1.3 EVAPOTRANSPIRATION DEFINITION

Evaporation and transpiration are not possible to specifically distinguish. Evaporation in a cropped soil is due to the solar radiation reaching the soil; during the growing period, evaporation decreases because of the increasing crop canopy. First, the water evaporates from the soil; then transpiration is the main process (Allen et al., 1998) (see Figure 1.1). Kool et al., (2014) stated transpiration depends on hydraulic resistance, root, leaf water potential, stomatal and leaf conductance. Temperature is based on atmospheric demands like soil water potential and hydraulic conductivity (Jones and Tardieu, 1998).

REFERENCES

Allen, R.G., Pereira, L.S., Howell, T.A., and Jensen, M.E. Evapotranspiration information reporting: I. Factors governing measurement accuracy. Agric. Water Manag., 98, (2011): 899–920.

Allen, R.G., Pereira, L.S., Raes, D., and Smith, M. Crop evapotranspiration: Guidelines for computing crop water requirements. In: United Nations FAO, Irrigation and Drainage Paper 56. FAO, Rome, Italy. (1998).

Amiri, M.J., Zarei, A.R., Abedi-Koupai, J., and Eslamian, S.T. The performance of fuzzy regression method for estimating of reference evapotranspiration under controlled environment. Int. J. Hydrol. Sci. Technol., 9, no 1, (2019): 28–38.

Blaney, H.F., and Criddle, W.D. Determining water requirements in irrigated areas from climatological and irrigation data. In: Soil Conservation Service Technical Paper 96, US. Department of Agriculture, Washington, USA. (1950).

Droogers, P., and Allen, R.G. Estimating reference evapotranspiration under inaccurate data conditions. Irrig. Drain. Syst., 16, (2002): 33–45.

Fernandes, C., Cora, J.E., and Araujo, J.A.C. Reference evapotranspiration estimation inside greenhouse. Sci. Agric., 60, no 3, (2003): 591–594.

Hargreaves, G.H., and Samani, Z.A. Reference crop evapotranspiration from temperature. Appl. Eng. Agric., 1, no 2, (1985): 96–99.

Jensen, M.E., and Haise, H.R. Estimating evapotranspiration from solar radiation. J. Irrig. Drain., 89, no 4, (1963): 15–41.

Jones, H.G., and Tardieu, F. Modelling water relations of horticultural crops: A review. Sci. Hortic., 74, (1998): 21–46.

Kamali, M.E., Nazari, R., Faridhosseini, A., Ansari, H., and Eslamian, S. The determination of reference evapotranspiration for spatial distribution mapping using geostatistics. Water Resour. Manage., 29, (2015): 3929–3940.

Kool, D., Agama, N., Lazarovitcha, N., Heitmanc, J.L., Sauerd, T.J., and Ben-Gal, A. A review of approaches for evapotranspiration partitioning. Agr. For. Meteorol., 184, (2014): 56–70.

Landeras, G., Ortiz-Barredo, A., and López, J.J. Comparison of artificial neural network models and empirical and semi-empirical equations for daily reference evapotranspiration estimation in the Basque Country (Northern Spain). Agric. Water Manag., 95, (2008): 553–565.

Li, Y., Qin, Y., and Rong, P. Evolution of potential evapotranspiration and its sensitivity to climate change based on the Thornthwaite, Hargreaves, and Penman—Monteith equation in environmental sensitive areas of China. Atmos. Res., 273, (2022): 106178.

Lingling, Z., Xia, J., Xu, C.-Y., Wang, Z., and Sobkowiak, L. Evapotranspiration estimation methods in hydrological models. J. Geogr. Sci., 23, no 2, (2013): 359–369.

Liu, C.M., and Zhang, D. Temporal and spatial change analysis of the sensitivity of potential evapotranspiration to meteorological influencing factors in China. Acta Geograph. Sin., 66, no 5, (2011): 579–588.

Ma, X.N., Zhang, M.J., Wang, S.J., Ma, Q., and Pan, S. Evaporation paradox in the Yellow River Basin. Acta Geograph. Sin., 67, no 5, (2012): 645–656.

Mpusia, P.T. Comparison of water consumption between greenhouse and outdoor cultivation. Ph.D. thesis, International Institute for Geo-Information Science and Earth Observation, The Netherland. (2006).

Pereira, L.S., Perrier, A., Allen, R.G., and Alves, I. Evapotranspiration: Concepts and future trends. J. Irrig. Drain. Eng., 125, no 2, (1999): 45–51.

Peterson, T.C., Golubev, V.S., and Groisman, P.Y. Evaporation losing its strength. Nature, 377, (1995): 687–688.

Rácz, C., Nagy, J., and Csaba, D.A. Comparison of several methods for calculation of reference evapotranspiration. Acta Silv. Lign. Hung., 9, (2013): 9–24.

Thornthwaite, C.W. An approach toward a rational classification of climate. Geogr. Rev., 38, no 1, (1948): 55–94.

Wilcox, B.P., Breshears, D.D., and Seyfried, M.S. Water balance on rangelands. In: Stewart, B.A., and Howell, T.A. (eds.). Encyclopedia of Water Science. Marcel Dekker, Inc., New York, (2003): 791–794.

Xu, C.Y., and Singh, V.P. Evaluation and generalization of temperature-based methods for calculating evaporation. Hydrol. Process., 15, no 2, (2000): 305–319.

Xu, Z.X., and Li, J.Y. Estimating basin evapotranspiration using distributed hydrologic model. J. Hydrol. Eng., 8, no 2, (2003): 74–80.

Zhao, C., Nan, Z., and Cheng, G. Methods for estimating irrigation needs of spring wheat in the middle Heihe basin. China. Agric. Water Manag., 75, (2005): 54–70.

2 Factors Affecting Evapotranspiration

It is possible to distinguish weather parameters, crop factors and other environmental factors that can affect ET. Among the weather parameters, we consider air radiation, air temperature, humidity and wind speed. Among the crop factors are crop type, crop height, crop roughness, reflection and ground cover. In particular, crop evaporation at standard conditions (ET_c) is the evaporating demand from crops in a large field with optimum soil water (Allen et al., 1998). Other factors that can affect ET are soil characteristics such as salinity and fertility, the use of fertilizer, the presence of pests and soil management and plant characteristics such as leaf anatomy, stomatal characteristics, aerodynamic properties and albedo. It is also necessary to consider ground cover, plant density and water content to avoid under- or overwatering. ET is distinguished as ET_0 (crop evaporation), ETc (crop evaporation at standard conditions) and ETc adj (crop evaporation at non-standard conditions); ETc adj is considered under variations in conditions such as pests and diseases, soil salinity, low soil fertility, and water shortage or waterlogging. A water stress coefficient K_s is calculated for the latter measurement, and Kc is used for all the other stresses (Allen et al., 1998). The U.N. Food and Agriculture Organization has calculated ET_0 for different agroclimatic regions (Allen et al., 1998) (Table 2.1).

Generally, the methods used to estimate ET show overestimates; for example, the modified Penman method overestimates ET_0 up to 20% under low evaporative conditions. Measuring data for estimating ET can contain systematic and random errors. Systematic errors include improper sensor function, operation, placement and

TABLE 2.1
Average ET_0 for Different Agroclimatic Regions in Millimeters per Day

	Mean daily temperature (°C)		
Regions	Cool (~10°C)	Moderate (20°C)	Warm (>30°C)
Tropics and subtropics	2–3	3–5	5–7
Humid and sub-humid	2–4	4–6	6-8
Arid and semiarid			
Temperate region	1–2	2–4	4–7
Humid and sub-humid	1–3	4–7	6–9
Arid and semiarid			

Source: Modified from Allen et al. (1998).

DOI: 10.1201/9781003467229-2

recording; non-representative vegetation and similar parameters. Random errors are due to components, electronic noises, vegetation and soil water management. There are also human errors such as in logging the data, assembling equipment or taking the actual measurements. Better training and experience could make them avoidable (Allen et al., 2011).

Tabari (2010) measured ET_0 in Iran with deficient data under cold humid, arid, warm humid and semiarid climatic conditions using four equations: Makkink (Makk), Turc (Tc), Priestley–Taylor (PT) and Hargreaves-Samani (HG). Tc accurately estimated daily and monthly ET_0; HG well estimated potential evapotranspiration (PET) in the semiarid regions, and Tc was effective in humid climate. These four equations were compared with FAO56PM to validate the results. Along the coastal regions, HG gave higher values than FAO56PM; it is possible that transmissivity affects many parameters like atmospheric moisture (humidity), so that the solar radiation at the surface is less attenuated. However, HG does not consider the attenuation of cloud cover, atmospheric moisture or atmospheric particulates (Fontenot, 2004).

Tabari (2010) found that HG underestimated ET_0 because of the high wind velocity. PT gave higher values than FAO56PM, whereas Makk showed lower values. Tc was developed for humid areas, but in warm humid climates PT well estimated ET_0 at the highest altitude; it gave very low values at the lowest altitudes. Tc showed good results at high altitude but also very low results at the lowest altitudes. In other warm humid climates, the Tc model performed very well, and HG, Makk and PT showed low values. In semiarid climates, HG performed well at estimating ET_0, whereas PT and Makk did not; PT gave the lowest values, lower than FAO56PM, Makk, Tc and HG. The equations performed the same in arid climates as in semiarid. Each of the ET_0 equations performs the best in the climate it was developed for.

REFERENCES

Allen, R.G., Pereira, L.S., Howell, T.A., and Jensen, M.E. Evapotranspiration information reporting: I. Factors governing measurement accuracy. Agric. Water Manag., 98, (2011): 899–920.
Allen, R.G., Pereira, L.S., Raes, D., and Smith, M. Crop evapotranspiration: Guidelines for computing crop water requirements. In: United Nations FAO, Irrigation and Drainage Paper 56. FAO, Rome, Italy. (1998).
Fontenot, R.L. An evaluation of reference evapotranspiration models in Louisiana. Louisiana State University and Agricultural & Mechanical College, Baton Rouge, LA, USA. (2004).
Tabari, H. Evaluation of reference crop evapotranspiration equations in various climates. Water Resour. Manag., 24, (2010): 2311–2337.

3 Evapotranspiration Estimation Methods

3.1 MICROCLIMATOLOGY

ET requires energy exchange at the vegetation level, and it lasts until the available energy is consumed, so it is fundamental for the principle of energy conservation. The equation for an evaporating surface is the following:

$$R_n - G - \lambda ET - H = 0 \tag{1}$$

R_n = net radiation; H = sensible heat; G = soil heat flux; λET = latent heat flux.

This equation must be applied to large surfaces and homogeneous vegetation because the vertical fluxes are considered but the energy transferred horizontally is not.

3.2 MASS TRANSFER

ET can be estimated using mass transfer, the vertical movement of small parcels of air (eddies) above a large homogenous surface. Eddies transport water vapor, heat and momentum from and to the evaporating surface. The ET is the vertical gradients of air temperature and water according to the Bowen ratio, assuming that eddy transfer coefficients for water vapor are proportional to those for heat and momentum. Bowen (1926) defined the ratio as follows:

$$\beta = \frac{H}{\lambda E} = \frac{\gamma \Delta T_a}{\Delta e} \tag{2}$$

β = dimensionless; H and λE = [W/m^2]; γ = psychometric constant [kPa/°C]; ΔT_a [°C] and Δe [kPa] = differences in ambient temperature and vapor pressure between two heights. Ashktorab et al., (1989) used the Bowen ratio–energy balance (BREB) to estimate ET and used micro BREB to measure evaporation close to the surface of a bare field):

$$\lambda E = \frac{R_n - G}{1 + \beta} \tag{3}$$

λE = energy flux density associated with the latent energy of evaporation; R_n = net radiation [W/m^2]; G = soil heat flux [w/m^2]; β = Bowen's ratio of sensible to latent heat flux.

This formula requires highly precise temperature and vapor measurements to determine the gradients between two heights.

DOI: 10.1201/9781003467229-3

3.3 EDDY COVARIANCE

Brutsaert (1982) defined eddy covariance as a measure of evaporation using momentum, temperature and water vapor:

$$ET = \rho w'q' \tag{4}$$

$ET = [kg/m^2\ s]$; ρ = air density $[kg/m^3]$; w' = mean covariance between vertical wind speed $[m/s]$ and q'= specific humidity $[kg/kg]$.

Baldocchi and Meyers, (1991) and Wilson et al., (2001) used this formula to measure evaporation under a forest canopy. This method requires a homogeneous surface without disturbance between the surface and the instrument, and the accuracy depends on the height of the instrument.

3.4 SOIL–WATER BALANCE

Measuring ET based on the soil–water balance means determining the incoming and outgoing water flux into the crop root zone at a certain time (Figure 3.1).

Fluxes such as subsurface flow, deep percolation and capillary rise are difficult to measure, and if time is short, they are not considered. Equation (5) gives good values for about seven to 10 days:

$$ET = I + P - RO - DP + CR \pm \Delta SF \pm \Delta SW \tag{5}$$

I = irrigation [mm], P = rainfall [mm], RO = surface runoff [mm], DP = deep percolation [mm], CR = capillary rise [mm/day], SF = subsurface flow [mm] and SW = variation soil water content [mm]. Irrigation and rainfall increase the water amount in soil, and

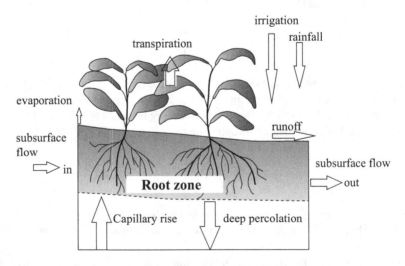

FIGURE 3.1   Soil–water balance in the root zone.

Source: Modified from Allen et al. (1998).

runoff and deep percolation decrease the amount. Capillary rise from a shallow water table moves towards the root zone or is transferred horizontally by subsurface flow (ΔSF).

According to Allen et al., (2011), this method is for large areas of land and water. The equation calculates seasonal average water evaporated and transpired from agricultural and non-agricultural areas. Precipitation is the main input therefore, predicted ET is reliable if the precipitation data are accurate.

3.5 LYSIMETERS

Computing evaporation with a lysimeter is difficult and expensive. When crops grown in a tank, water is lost by changes in mass, and the lysimeter gives a very accurate ET measurement. However, a disadvantage of the method is that the vegetation inside and outside the lysimeter must be of the same height and leaf area. Unfortunately, this characteristic has often resulted in inaccurate ET_c and K_c values. Allen et al., (2011) affirmed that this method is an old one but confirmed that if the data concerning vegetation and environmental conditions are respected, the values are reliable given that lysimeters are sensitive to environmental factors (Allen et al., 2011).

There are three different types of lysimeters: non-weighing with a constant water table, non-weighing percolation and weighing. The weighing type gives good results when it is well managed for a short time. Allen et al., (2011) highlighted that a typical mistake with using lysimeters is inputting the incorrect evaporating area. A further disadvantage is the canopy area that may exceeds the lysimeter area, particularly if the vegetation outside the lysimeter is different from that one inside this can increase the solar radiation interception.

Armanios et al., (2000) selected 12 regression equations to estimate ET_0 and compare the results with lysimeter data in the area of Wadi El-Natroon on the Cairo-Alexandria desert highway: Penman (1963), Penman–Monteith (FAO56PM), FAO-24 corrected Penman, Penman–Monteith (daytime wind), Jensen–Haise (JH), FAO-24 radiation, Priestley-Taylor (PT), Priestley–Taylor ($\alpha = 1$), SCS Blaney–Criddle, FAO-24 Blaney–Criddle, Hargreaves (HG) and FAO-24 Pan. Frevert et al., (1983) used FAO-24 correct Penman, FAO-24 Radiation, FAO-24 Blaney–Criddle and FAO-24 Pan to calculate the coefficients. See Table 3.1 for the classifications of the equations and Table 3.2 for the ranks and statistics by period for the daily ET_0 estimates.

These equations are a combination of theory and empirical formulation based on solar radiation, temperature and pan evaporation, and the reference crop that Armanios et al., (2000) used was alfalfa (*Medicago sativa*). They calculated the standard error of estimate (SEE) between the calculated ET_0 and the daily lysimeter value as follows:

$$\text{SEE} = \sqrt{\frac{\sum_{i=1}^{n}(Y_i - Y)^2}{n-1}} \qquad (6)$$

Y_i = the average i^{th} average daily lysimeter ET_0; Y = the corresponding calculated ET_0 and n = the total number of observations.

TABLE 3.1
Equations and their Classifications

Classification	Equation
Combination	Penman (1963), Penman–Monteith, FAO-24 correct Penman, Penman–Monteith (daytime wind)
Radiation	JH, FAO-24 radiation, PT, Priestley–Taylor ($\alpha = 1$)
Temperature	SCS Blaney–Criddle, FAO-24 Blaney–Criddle, HG
Pan evaporation	FAO-24 Pan

Source: Modified from Armanios et al. (2000).

TABLE 3.2
Rank, Period and Statistics for the Daily Estimates of ET_0

Rank	Method	All Period				Peak period			
		SEE	b	R	ASEEs	SEE	b	R	ASEE
1	FAO56PM	0.727	1.026	0.917	0.713	0.847	1.06	0.99	**0.713**
2	Original Penman	0.831	0.92	0.926	0.674	0.848	0.954	0.989	**0.752**
3	Hargreaves	0.832	1.023	0.89	0.823	0.993	1.064	0.985	**0.832**
4	SCS Blaney–Criddle	0.913	0.974	0.868	0.901	0946	0.991	0.982	**0.943**
5	Jensen–Haise	0.918	0.973	0.867	0.905	0.997	0.991	0.98	**0.994**
6	Penman daytime wind	1.177	0.867	0.892	0.815	1.051	0.913	0.99	**0.71**
7	Priestley–Taylor	1.258	1.215	0.899	0.788	1.538	1.182	0.984	**0.904**
8	FAO 24-pan evaporation	1.379	1.012	0.691	1.377	1.313	1.097	0.976	**1.102**
9	FAO radiation	1.623	0.798	0.893	0.812	1.946	0.822	0.985	**0.857**
10	FAO Blaney–Criddle	1.773	0.832	0.915	0.724	1.773	0.835	0.988	**0.769**
11	FAO Penman	1.715	0.778	0.934	0.636	2.313	0.787	0.989	**0.742**
12	Priestley–Taylor ($\alpha = 1$)	1.719	1.358	0.864	0.915	2.803	1.484	0.981	**0.975**

Source: Modified from Armanios et al. (2000).

The SEE refers to how closely each method approaches the lysimeter ET_0 values. Armanios et al., (2000) performed linear regression analyses to compare the mean SEEs from each equation with the lysimeter values using the average daily lysimeter ET_0 as the dependent variable and the calculated ET_0 values as the independent variable. The linear regression is represented by Equation (7):

$$ET_0 \text{ (Lysimeter)} = bET_0 \text{ (Calculated)} \tag{7}$$

b = adjusted standard error of estimation (ASEE), which is calculated by multiplying ET_0 by b and the new SEE.

The authors determined that FAO56PM calculated the most reliable ET_0 for the studied area. Some other equations showed systematic errors due to the high corrections on their SEEs as b is adjusted (Armanios et al., 2000).

A micro-lysimeter (ML) is placed on the soil surface and sealed. The weight difference from the beginning to the desired end point is proportional to the evaporation according to Shawcroft and Gardner, (1983), Walker (1983) and Armanios et al., (2000). Boats and Robertson, (1982) found that ML gave reliable results for 24–48 hours without great differences in field conditions. Kool et al., (2014) pointed out that MLs can be used to validate other methods.

3.6 PAN EVAPORATION

Evaporation from a pan, an open water surface, considers parameters such as air temperature, air humidity, wind, radiation and evapotranspiration. This method has been used to apply empirical coefficients to relate pan evaporation to ET_0. This method is suitable for 10 days or longer. Microclimatic conditions and station maintenance factors affect this method. During dry periods, the evaporated water is estimated based on the decreasing water depth.

REFERENCES

Allen, R.G., Pereira, L.S., Howell, T.A., and Jensen, M.E. Evapotranspiration information reporting: I. Factors governing measurement accuracy. Agric. Water Manag., 98, (2011): 899–920.

Armanios, S., El Quosy, D., Abdel Monim, Y.K., and Gaafar, I. Evaluation of evapotranspiration equations using weighing Lysimeter data. Sci. Bull., 35, no 1, (2000): 193–205.

Baldocchi, D.D., and Meyers, P.T. Trace gas exchange above the floor of a deciduous forest 1: Evaporation and CO_2 efflux. J. Geophys. Res., 96, (1991): 7271–7285.

Boast, C.W., and Robertson, T.M. A micro-lysimeter method for determining evaporation from bare soil: Description and laboratory evaluation. Soil Sci. Soc. Am. J., 46, (1982): 689–696.

Bowen, I.S. The ratio of heat losses by conduction and by evaporation from any water surface. Phys. Rev., 27, no 6, (1926): 779–787.

Brutsaert, W. Evaporation Into the Atmosphere: Theory, History and Applications. D. Reidel Publishing, Dordrecht, The Netherlands, (1982).

Frevert, D.K., Hill, R.W., and Braaten, B.C. Estimation of FAO evapotranspiration coefficients. J. Irri. Drain. ASCE, 109, no IR2, (1983): 265–270.

Kool, D., Agama, N., Lazarovitcha, N., Heitmanc, J.L., Sauerd, T.J., and Ben-Gal, A. A review of approaches for evapotranspiration partitioning. Agr. For. Meteorol., 184, (2014): 56–70.

Penman, H.L.Vegetation and hydrology. Soil Sci., 96, no 5, (1963): 357.

Shawcroft, R.W., and Gardner, H.R. Direct evaporation from soil under a row crop canopy. Agric. Meteorol., 28, (1983): 229–238.

Walker, G.K. Measurement of evaporation from soil beneath crop canopies. Can. J. Soil Sci., 63, (1983): 137–141.

Wilson, K.B., Hanson, P.J., Mulholland, P.J., Baldocchi, D.D., and Wullschleger, S.D. A comparison of methods for determining forest evapotranspiration and its components: Sap-flow, soil water budget, eddy covariance and catchment water balance. Agric. For. Meteorol., 106, (2001): 153–168.

4 Empirical Methods
Some Examples

4.1 MODIFIED PENMAN METHOD

This method gives the best results with minimum error using grass as the reference crop (Allen et al., 1998). This method is valid for 10 days up to a month, and it is recommended that practitioners make local wind calibrations for good results.

4.2 BLANEY–CRIDDLE METHOD

Blaney–Criddle (BC) is recommended for periods of a month or longer when climate data such as air temperature are easily available:

$$ET_0 = a + b[P(0.46T + 8.13)] \tag{8}$$

T = mean monthly temperature [°C]; P = mean annual percentage of daytime hours (Doorenbos and Pruitt, 1977) and a, b = climate factors.

4.3 PENMAN–MONTEITH METHOD

The Penman–Monteith equation (FAO56PM) calculates the parameters for the exchange energy and latent heat flux (evapotranspiration) from uniform vegetation. Most of these parameters can be calculated from the data recorded from weather stations and/or agrometeorological stations posted in cropped areas, they are affected by the atmospheric conditions, the height for these stations is 2 m above the surface. Moreover, the data from weather stations must be validated before being applied in FAO56PM (see Equation (9)). Allen et al., (2011) suggested not measuring ET from small or medium-sized patches of vegetation or clusters of small fields to avoid the clothesline effect. The air–vegetation interchange is highly efficient, in contrast with large vegetation, and ET can be higher than ETref. In arid and humid climates, this method generally achieves accurate results.

$$\lambda ET = \frac{\Delta\left(R_n - G\right) + \rho_a c_p \dfrac{\left(e_s - e_a\right)}{r_a}}{\Delta + \gamma\left(1 + \dfrac{r_s}{r_a}\right)} \tag{9}$$

R_n = net radiation, G = soil heat flux, $(e_s - e_a)$ = vapor pressure deficit of the air, ρ_a = mean air density (constant pressure), c_p = specific heat of the air, Δ = slope of the saturation vapor pressure–temperature relationship, γ = psychrometric constant, r_s and r_a = (bulk) surface and aerodynamic resistance.

DOI: 10.1201/9781003467229-4

This equation was improved by introducing the resistance r_s and r_a to extend it to cropped surface, where r_s is the resistance of vapor flow through stomata openings, total leaf area and soil surface, r_a is the resistance from the vegetation upward involving friction from air flowing over surfaces (Figure 4.1, Equation (10)) (Allen et al., 1998).

$$r_a = \frac{\ln\left[\dfrac{z_m - d}{z_{om}}\right] ln\left[\dfrac{z_h - d}{z_{oh}}\right]}{k^2 u_z} \tag{10}$$

r_a = aerodynamic resistance [s/m]; z_m = height of wind measurement [m]; z_h = height of humidity measurement [m]; d = zero plane displacement height [m]; z_{om} = roughness length governing momentum transfer; z_{oh} = roughness length governing momentum transfer of heat and vapor [m]; k = von Karman's constant, 0.41; u_z = wind speed at height z [m/s].

This equation is valid for under stable temperature, atmospheric pressure and wind velocity in adiabatic conditions (no heat exchange).

The (bulk) surface resistance is the resistance of vapor flow through the transpiring crop and evaporating soil surface (see Figure 4.1) and is calculated as follows:

$$r_s = \frac{r_l}{LAI_{active}} \tag{11}$$

r_s = (bulk) surface resistance [s/m]; r_l = bulk stomatal resistance of the well-illuminated leaf [s/m]; LAI_{active} = active (sunlit) leaf area index [m² (leaf area)/m² (soil surface)]. The leaf area index (LAI) is the upper area of leaves; LAI_{active} is the leaf area that actively contributes to surface heat and vapor transfer; values of 3–5 are valid for many mature crops depending on plant density and crop variety.

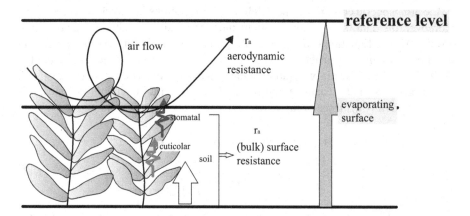

FIGURE 4.1 Simplified representation of the (bulk) surface and aerodynamic resistances for water vapor flow.

Source: Modified from Allen et al. (1998).

The bulk stomatal resistance is the resistance of a leaf. It is crop specific and differs according to crop variety and management. The resistance increases as the crop ripens depending on climate and water availability (see Figure 4.2). Unfortunately, there is a lack of information on the changes according to the type of crop. Bulk stomatal resistance increases because there are water deficiencies in the crop and the soil water limits the crop ET.

As noted, Allen et al., (1998) used grass as the reference crop, but grass variety and morphology deeply affect ET, and there are also differences between warm and cool seasons. Allen et al., (1998) suggests a hypothetical reference crop with an assumed crop height of 0.12 m, a fixed surface resistance of 70 s/m and an albedo of 0.23 (see Figure 4.3).

One important assumption is that all the fluxes are one-dimensional upwards and that the reference surface is a green grass of uniform height, with shaded ground and adequate water.

In 1990, FAO56PM was updated to consider the aerodynamic features and surface resistance. This equation determines ET considering a hypothetical grass reference surface, and the obtained values are valid in different periods of the year and in different regions. These values can also be compared with the inferred data for other crops.

$$ET_0 = \frac{0.408\Delta\left(R_n - G\right) + \gamma\dfrac{900}{T+273}u_2\left(e_s - e_a\right)}{\Delta + \gamma\left(1 + 0.34u_2\right)} \tag{12}$$

ET_0 = reference ET [mm/day]; R_n = net radiation at the crop surface [MJ/m^{2}day]; G = soil heat flux density [MJ/m^{2}day]; T = mean daily air temperature at 2 m height

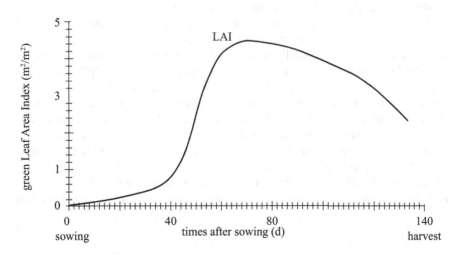

FIGURE 4.2 Active LAI over the growing season for a maize crop.

Source: Modified from Allen et al. (1998).

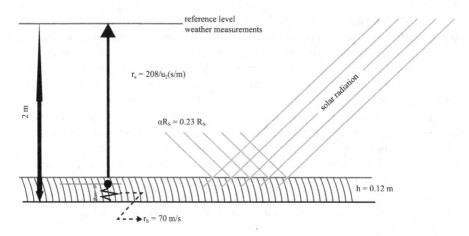

FIGURE 4.3   Characteristics of the hypothetical reference crop.

Source: Modified from Allen et al. (1998).

[°C]; u_2 = wind speed at 2 m height [m/s]; e_s = saturation vapor pressure [kPa]; e_a = actual vapor pressure [kPa]; e_s-e_a = saturation vapor pressure deficit [kPa]; Δ = slope of vapor pressure curve [kPa/°C]; γ = psychrometric constant [kPa/°C].

For the accuracy of results found with Equation (12), it is important that the weather measurements have been done at 2 m above an extensive surface of green grass, shading the ground and without shortage of water. The disadvantage is that errors in the values are due to errors in the measurements; moreover, local environmental and management factors could affect ET_0 observations.

To compute FAO56PM equation, it is necessary to know the air temperature, humidity, radiation, wind speed, altitude and latitude to estimate local atmospheric pressure and extraterrestrial radiation (Allen et al., 1998). The procedures for estimating missing data should be validated at the regional level (Allen et al., 1998). If only the mean daily temperature is available, ET_0 could be underestimated because of a lower saturation vapor pressure (e_s) and lower vapor pressure difference ($e_s - e_a$); the saturation vapor pressure–temperature relationship becomes nonlinear (Figure 4.4). The daily vapor pressure can be inferred from the maximum and minimum relative humidity, psychrometric constant or dewpoint temperature. To achieve better results in the equation, mean daily maximum and minimum air temperatures (°C) should be used when they are available. The humidity (kPa) is necessary for the calculation of ET_0. The relative humidity (RH) expresses the degree of saturation of the air as follows (see Figure 4.5):

$$RH = 100 \frac{e_a}{e^o\left(T\right)} \qquad (13)$$

e_a = water vapor in the air; e_o = saturation water vapor at the same temperature.

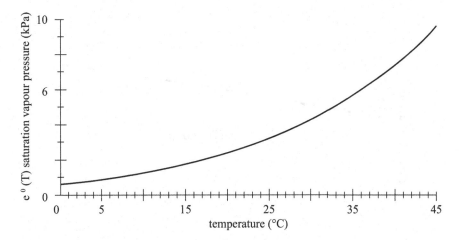

FIGURE 4.4 Saturation vapor pressure as a function of temperature: e°(T) curve.
Source: Modified from Allen et al. (1998).

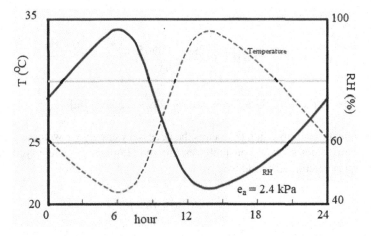

FIGURE 4.5 Relative humidity over 24 hours for a constant actual vapor pressure of 2.4 kPa.
Source: Modified from Allen et al. (1998).

The daily net radiation (MJ/m^2day) can be derived from the short-wave radiation; it can be measured with a pyranometer or calculated from the daily duration of bright sunshine measured with a Campbell–Stokes sunshine recorder (Allen et al., 1998). The potential amount of radiation depends on latitude and season. For wind speed, it is critical to measure it at the height of 2 m; other heights give inaccurate wind speeds. ET also depends on wind and air turbulence, because the large amount of air is transferred over the evaporating surface (Figure 4.6).

The atmospheric pressure exerts a small effect on ET_0 calculation, so the average local value is sufficient, calculated as follows:

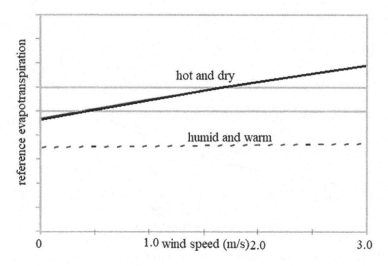

FIGURE 4.6   The effects of wind speed on evapotranspiration in hot, dry and warm, humid weather.

Source: Modified from Allen et al. (1998).

$$P = 101.3\left(\frac{293 - 0.0065z}{293}\right)^{5.26} \tag{14}$$

P = atmospheric pressure [kPa]; z= elevation above sea level [m].

The latent heat of vaporization (λ) means the necessary energy for water to pass from liquid state to vapor state, at constant pressure and temperature. The psychrometric constant (γ) is calculated as follows:

$$\gamma = \frac{Pc_p}{\varepsilon\lambda} = 0.665x10^{-3}P \tag{15}$$

γ = psychrometric constant [kPa/°C], P = atmospheric pressure [kPa]; λ = latent heat of vaporization, 2.45 [MJ/kg]; c_p = specific heat at constant pressure, 1.013 × 10⁻³ [MJ/kg°C]; ε = ratio molecular weight of water vapor/dry air = 0.622.

The specific heat is the energy to increase the temperature of a unit mass of air by one degree at constant pressure. The saturation vapor pressure is related to air temperature (Figure 4.6).

$$e^o = 0.6108exp\left[\frac{17.27T}{T + 237.3}\right] \tag{16}$$

e° = saturation vapor pressure at T [kPa]; T = air temperature [°C]; exp = 2.7183 (base of natural logarithm) raised to the power.

The mean saturation vapor pressure should be the mean of the pressure at the mean daily maximum and minimum temperatures for that period, otherwise, the values will be underestimated; hence, it is suggested to calculate the saturation vapor pressure at minimum and maximum air temperature:

$$e_s = \frac{e^o\left(T_{max}\right) + e^o\left(T_{min}\right)}{2} \tag{17}$$

The slope between saturation pressure and temperature is calculated as follows:

$$\Delta = \frac{4098\left[0.6108\exp\left(\dfrac{17.27T}{T+237.3}\right)\right]}{\left(T+273.3\right)^2} \tag{18}$$

Δ = slope saturation vapor pressure curve at T [kPa/°C]; T = air temperature [°C]; exp[. . .]2.7183 (base of natural logarithm) raised to the power.

In FAO56PM, Δ is calculated using mean air temperature $((T_{max} + T_{min})/2)$. The vapor pressure is calculated from the difference between the dry and wet bulb temperatures. It can be determined using the dew point temperature, derived from psychrometric data and relative humidity data:

$$e_a = e^o\left(T_{dew}\right) = 0.6108\exp\left[\frac{17.27T_{dew}}{T_{dew}+237.3}\right] \tag{19}$$

$$e_a = e^o\left(T_{wet}\right) - \gamma_{psy}\left(T_{dry} - T_{wet}\right) \tag{20}$$

e_a = actual vapor pressure [kPa]; e^o (T_{wet}) = saturation vapor pressure at wet bulb temperature [kPa]; γ_{psy} = psychrometric constant of the instruments [kPa/°C]; $T_{dry} - T_{wet}$ = wet bulb depression, where T_{dry} is the dry bulb and T_{wet} is the wet bulb temperature [°C]:

$$\gamma_{psy} = a_{psy}P \tag{21}$$

a_{psy} = coefficient depending on the type of ventilation of the wet bulb [°C⁻¹] and P= atmospheric pressure [kPa]. The coefficient a_{psy} depends mainly on the design of the psychrometer and the rate of ventilation around the wet bulb:

$$e_a = \frac{e^o\left(T_{min}\right)\dfrac{RH_{max}}{100} + e^o\left(T_{max}\right)\dfrac{RH_{min}}{100}}{2} \tag{22}$$

e_a = actual vapor pressure [kPa]; $e^o(T_{min})$ = saturation vapor pressure at daily minimum temperature [kPa]; $e^o(T_{max})$ = saturation vapor pressure at daily maximum temperature [kPa]; RH_{max} = maximum relative humidity [%]; RH_{min} = minimum relative humidity [%].

If there are errors in RH_{min}, it is suggested to use RH_{max}:

$$e_a = e^o\left(T_{min}\right)\frac{RH_{max}}{100} \tag{23}$$

In the absence of both RH_{max} and RH_{min}, e_a can be calculated as follows:

$$e_a = \frac{RH_{mean}}{100}\left[\frac{e^o\left(T_{max}\right)+e^o\left(T_{min}\right)}{2}\right] \tag{24}$$

RH_{mean} is the mean relative humidity, defined as the average between RH_{max} and RH_{min}.

If data about humidity are dubious, the actual vapor pressure can be determined considering the dew point temperature close to the minimum temperature:

$$e_a = e^o\left(T_{min}\right) = 0.611 exp\left[\frac{17.27T_{min}}{T_{min}+237.3}\right] \tag{25}$$

The radiation reaching a horizontal plane is known as solar radiation, Rs, and as it penetrates the atmosphere, some of the radiation is scattered, reflected or absorbed by the atmospheric gases, clouds and dust (Figure 4.7).

The net radiation (Rn) is the balance between the energy absorbed, reflected and emitted by the earth's surface or the difference $R_{ns} - R_{nl}$ (incoming radiation shortwave and outgoing radiation longwave. Rn is generally positive in daytime, not

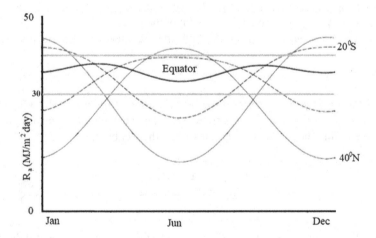

FIGURE 4.7   Annual variation in extraterrestrial radiation (Ra) at the equator, 20° and 40° north and south.

Source: Modified from Allen et al. (1998).

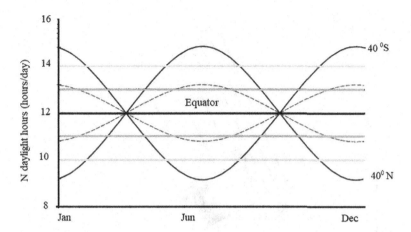

FIGURE 4.8 Annual variation in the daylight hours (N) at the equator and at 20 and 40° north and south.

Source: Modified from Allen et al. (1998).

at high latitudes. If data about longwave and net radiation are lacking, they could be inferred from the weather parameters. Soil heat flux G is the energy required to heat the soil, and it is positive when the soil is warming and negative when the soil is cooling. See Figure 4.8 for annual variations in daylight hours depending on location.

Solar radiation is measured using pyranometers, radiometers or solarimeters. If these instruments are not available, it is estimated according to the duration of bright sunshine measured with the Campbell–Stokes sunshine recorder (Allen et al., 1998). The extraterrestrial radiation (Ra) for each day and according to the latitudes could be estimated from the solar constant, the solar declination and time of the year (Figure 4.9, Equation 26):

$$R_a = \frac{24(60)}{\pi} G_{sc} d_r [\omega_s \sin\varphi \sin\delta + \cos\varphi \cos\delta \sin\omega_s] \qquad (26)$$

R_a = extraterrestrial radiation [MJ/m²day]; G_{sc} = solar constant = 0.0820 MJ/m²min; d_r = inverse relative distance from Earth to the Sun; ω_s = sunset hour angle [rad]; ϕ = latitude [rad]; δ = solar declination angle [rad]; d_r and δ are calculated as

$$d_r = 1 + 0.033\cos\left(\frac{2\pi}{365} J\right) \qquad (27)$$

$$\delta = 0.409\sin\left(\frac{2\pi}{365} J - 1.39\right) \qquad (28)$$

J = number of the day in the year between 1 (1 January) and 365 or 366 (31 December).

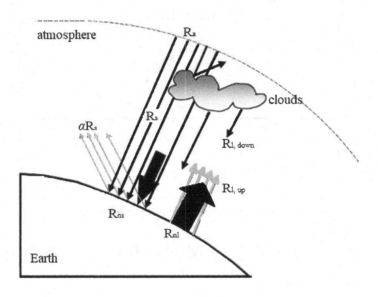

FIGURE 4.9   Various components of radiation.

Source: Modified from Allen et al. (1998).

$$\omega_s = arccos\left[-tan\varphi tan\delta\right] \tag{29}$$

$$\omega_s = \frac{\pi}{2} - arctan\left[\frac{-tan\varphi tan\delta}{X^{0.5}}\right] \tag{30}$$

$$X = 1 - [\tan\varphi]^2[\tan\delta]^2 \tag{31}$$

X = 0.00001 if X ≤ 0.

Solar radiation can be calculated with the Ångstrom formula:

$$R_s = (a_s + b_s\frac{n}{N})R_a \tag{32}$$

Rs = solar or shortwave radiation [MJ/m²day]; n = actual duration of sunshine [hour]; N = maximum possible duration of sunshine or daylight hours [hour]; n/N = relative sunshine duration; Ra = extraterrestrial radiation [MJ/m²day]; a_s = regression constant expressing the fraction of terrestrial radiation reaching the earth on overcast days (n = 0); $a_s + b_s$ = fraction of extraterrestrial radiation reaching the earth on clear days (n = N).

Solar radiation measurements are less valid over time in mountain or coastal regions because exposure, altitude and variable rainfall all influence the results. The values are most accuracy under clear skies when solar radiation is high because the atmosphere is transparent; the temperature is low during the night because this radiation is absorbed from the atmosphere. The difference between the temperatures indicates the extraterrestrial radiation (Figure 4.10) and is calculated using Hargreaves' radiation formula:

$$R_s = k_{Rs}\sqrt{\left(T_{max} - T_{min}\right)}R_a \tag{33}$$

R_a = extraterrestrial radiation [MJ/m²d]; T_{max} = maximum temperature [°C]; T_{min} = minimum air temperature [°C]; k_{Rs} = adjustment coefficient 0.16 ... 0.19) [°C$^{-0.5}$].

Net radiation is the difference between incoming shortwave (Rns) and outgoing net longwave radiation (Rnl):

$$R_n = R_{ns} - R_{nl} \tag{34}$$

The soil heat flux is lower than Rn, in particular if the soil is vegetated and the calculation time is about 24 hr or longer:

$$G = c_s \frac{T_i - T_{i-1}}{\Delta t} \Delta z \tag{35}$$

G = soil heat flux [MJ/m²day]; c_s = soil heat capacity [MJ/m³°C]; T_i = air temperature at time i [°C]; T_{i-1} = air temperature at time i-1[°C]; Δt = length of time interval [day]; Δz = effective soil depth.

Wind speed greatly affects the equation for evapotranspiration; it is important the height and anemometers are at 2 m above the surface (Figure 4.7 and Figure 4.11):

$$u_2 = u_z \frac{4.87}{ln\left(67.8z - 5.42\right)} \tag{36}$$

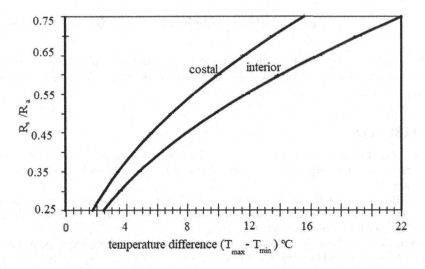

FIGURE 4.10   The relationships between extraterrestrial radiation reaching Earth's surface (R_s/R_a) and the difference of temperature $T_{max} - T_{min}$ for interior (K_{Rs} = 0.16) and coastal (K_{Rs} = 0.19) areas.

Source: Modified from Allen et al. (1998).

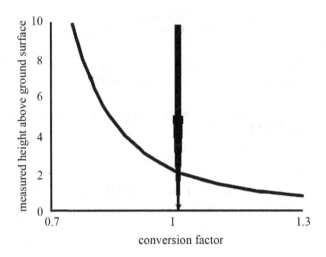

FIGURE 4.11 Converting wind speed measured at different heights above ground level to wind speed at the standard height (2 m).

Source: Modified from Allen et al. (1998).

u_2 = wind speed at 2 m above ground surface [m/s]; u_z = measured wind speed at z m above the ground surface [m/s]; z = height of measurement above ground surface [m]. Wind speed data are valid for long periods if the region is homogenous. Imported data are valid for monthly estimates of ET.

If weather data are missing like solar radiation, relative humidity and wind speed, it is possible to use the Hargreaves equation:

$$ET_0 = 0.0023\left(T_{mean} + 17.8\right)\left(T_{max} - T_{min}\right)^{0.5} R_a \qquad (37)$$

This equation must be verified for new regions and compared with FAO56PM close to the weather station to obtain valid solar radiation, air temperature, humidity and wind speed data.

REFERENCES

Allen, R.G., Pereira, L.S., Howell, T.A., and Jensen, M.E. Evapotranspiration information reporting: I. Factors governing measurement accuracy. Agric. Water Manag., 98, (2011): 899–920.

Allen, R.G., Pereira, L.S., Raes, D., and Smith, M. Crop evapotranspiration: Guidelines for computing crop water requirements. In: United Nations FAO, Irrigation and Drainage Paper 56. FAO, Rome, Italy. (1998).

Doorenbos, J., and Pruitt, W.O. Guidelines for predicting crop water requirements. In: FAO, UN, Irrigation and Drainage Paper No.24., 2nd ed., FAO, Rome, Italy. (1977).

5 Crop Coefficient

Allen et al., (2011) confirmed the limit of ET from FAO56PM due to the aerodynamic transport and equilibrium forces across the vegetation canopy. The upper limit is represented by ASCE-EWRI and the high reference values. This upper limit for crop evapotranspiration (ETc) is calculated by comparing ETref with a crop coefficient (Kc). The authors used two types of crops as references: cool-season grass or tall alfalfa (common symbols are ET_0 and ETr,); they list coefficients for the most used vegetation species (Allen et al., 1998, 2007a, 2011). The crop coefficient was defined in 1968 by Jensen. It is possible to express Kc as a ratio between ETc and Etref:

$$K_c = \frac{ET_c}{ET_{ref}} = \frac{\left[\Delta_c \left(R_{nc} - G_c \right) + \frac{\rho_a c_p (e_z^o - e_z)_c / r_{ac}}{r_{ac}} \right] / \left[\Delta_c + \gamma \left(1 + \frac{r_{sc}}{r_{ac}} \right) \right]}{\left[\Delta_r \left(R_{nr} - G_r \right) + \rho_a c_p (e_z^o - e_z)_r / r_{ar} \right] / \left[\Delta_r + \gamma \left(1 + \frac{r_{sr}}{r_{ar}} \right) \right]} \quad (38)$$

Rn = net radiation; G = soil heat flux density; ρ_a = air density; c_p = specific heat of air; e_z^o = saturation vapor pressure at z height; e_z = actual vapor pressure at z height; r_a = aerodynamic resistance to heat and vapor transport from surface to z height; Δ = slope of saturation vapor pressure–temperature curve; γ = psychrometric constant; r_s = bulk surface resistance.

The $_c$ in the parameters represents characteristic values for the actual vegetation, and $_r$ represents the same for the reference vegetation, grass or alfalfa. Allen et al., (2011) established that Kc depends on the roughness, leaf area and albedo of the vegetative surface and highlighted that the equilibrium on the surface level is conditioned by the difference between ET_c and ET_{ref}. Changing values for e_z^o, e_z, Δ and wind speed affect cooling and humidification and produce inaccurate values for Kc.

Allen et al., (2011) considered two reference surfaces: a 0.12 tall cool-season clipped grass according FAO56PM and a 0.5 m tall full-cover alfalfa. The latter features low surface resistance and quite high aerodynamic roughness to represent a large expanse of covered water. This kind of vegetation (alfa-alfa) shows a maximum ET of 1.0, showing the highest conversion of obtainable energy into latent heat flux (λE) (Allen et al., 2011). The grass reference showed a Kc of 1.3 in windy and semiarid conditions. In humid conditions, ET mostly depends on net radiation, not aerodynamics and vapor deficit (Allen et al., 2011). Reference crop (grass and alfalfa) shows the same albedo, therefore the energy, heat, vapor transfer rate, ET and Kc are more similar in humid and semi-humid conditions than in arid conditions.

Potential evapotranspiration (PET) is an index of hydrological budget, and it is fundamental in estimating regional biological processes (Lu et al., 2005) as well as

DOI: 10.1201/9781003467229-5

for measuring actual evapotranspiration (AET) in rainfall-runoff modeling. Rácz et al., (2013) pointed out that PET and AET must both be determined, and ET_0 was the reference evapotranspiration for a green grass surface of the same height, with ground shade and suitable water (Allen et al., 1998).

PET is an index of the amount of water loss in the atmosphere (Lu et al., 2005). According to Grismer et al., (2002) and Lu et al., (2005), PET can be estimated through lysimeters, generally by equations. There are many equations for calculating PET, but they unfortunately are specific to specific regions. Federer et al., (1996) and Lu et al., (2005) reported that these equations can give differences of hundreds of millimeters in their results. Moreover, the monthly water balance is sensitive to the applied equation. The authors recommend validating the equations in the field to choose the reliable PET equation for estimating AET.

Lu et al., (2005) compared three temperature-based methods, Thornthwaite (1946), Hamon (1963) and Hargreaves and Samani (1985), and three radiation methods, Turc (1961), Makkink (1957) and Priestley and Taylor (1972), both in a forest. They compared monthly and daily values acquired using annual PET to calculate AET.

The Thornthwaite method (1948) is given as follows:

$$PET = 1.6 L_d (\frac{10T}{I})^a \qquad (39)$$

PET = monthly PET [cm]; L_d = daytime length from sunrise to sunset in multiples of 12 hours; T = monthly mean air temperature [°C]; $a = 6.75 \times 10^{-7} I^3 - 7.71 \times 10^{-5} I^2 + 0.01791 I + 0.49239$; I = annual heat index, which is computed from the monthly heat indices.

The Hamon method (1963) (PET = 0 when T < 0) is as follows:

$$PET = 0.1651 \times L_d \times RHOSAT \times KPEC \qquad (40)$$

ETP = daily ETP [mm]; L_d = daytime length; RHOSAT= saturated vapor density [g/m³] at the daily mean air temperature [T].

$$RHOSAT = 216.7 \times \frac{ESAT}{(T + 273.3)} \qquad (41)$$

$$ESAT = 6.108 \times EXP \left(17.26939 \times \frac{T}{(T + 237.3)} \right) \qquad (42)$$

T = daily mean air temperature [°C]; ESAT = saturated vapor pressure [mb] at the given T; KPEC = calibration coefficient, which is set to 1.2 in Lu et al., (2005). The Turc method (1961) is as follows:

RH < 50%

$$PET = 0.013 \left(\frac{T}{T + 15} \right) (R_s + 50) \left(1 + \frac{50 - RH}{70} \right) \qquad (43)$$

$$RH > 50\%$$

$$PET = 0.013\left(\frac{T}{T+15}\right)(R_s + 50) \qquad (44)$$

PET = daily PET [mm/day]; T = daily mean air temperature [°C]; Rs = daily solar radiation [ly/day or cal/cm²/d] where cal/cm²/d = (100/4.1868) [MJ/m²/day]; RH = daily mean relative humidity [percent].

The Priestley–Taylor equation (1972) is as follows:

$$\lambda PET = \alpha \frac{\Delta}{\Delta + \gamma}(R_n - G) \qquad (45)$$

PET = daily PET [mm/day]; λ = latent heat of vaporization [MJ/kg] λ = 2.501 – 0.002361 T; T = daily mean air temperature [°C]; α = calibration constant; α = 1.26 for wet or humid conditions; Δ = slope of the saturation vapor pressure temperature curve [kPa/°C] Δ = 0.200 (0.00738 T + 0.8072)^7 – 0.000116; γ = psychrometric constant modified by the ratio of canopy resistance to atmospheric resistance [kPa/°C]:

$$\gamma = \frac{c_p p}{0.622 \lambda} \qquad (46)$$

c_p = specific heat of moist air at constant pressure [kJ/kg/°C] c_p = 1.013 [kJ/kg/°C] = 0.001013 [MJ/kg/°C]; p = atmospheric pressure [kPa] p = 101.3 – 0.01055 EL; EL = elevation [m]; Rn = net radiation [MJ/m²/day]; G = heat flux density to the ground [MJ/m²/day].

$$G = 4.2\frac{(T_{i+1} - T_{i-1})}{\Delta t} = -4.2\frac{(T_{i-1} - T_{i+1})}{\Delta t} \qquad (47)$$

T_i = mean air temperature [°C] for period i; Δt = difference in time (days) between two periods.

The Makkink method (1957) is given as follows:

$$PET = 0.61\left(\frac{\Delta}{\Delta + \gamma}\right)\frac{R_S}{58.5} - 0.12 \qquad (48)$$

Variables and units in this equation are the same as those for PT and Turc method. Hargreaves-Samani (HS) (1985):

$$\lambda PET = 0.0023 R_a T D^{0.5}(T + 17.8) \qquad (49)$$

PET = daily PET [mm/day]; λ = latent heat of vaporization [MJ/kg]; T = daily mean air temperature [°C]; R_a = extraterrestrial solar radiation [MJ/m²/day]; TD = daily difference between the maximum and minimum air temperature [°C].

Lu et al., (2005) made some assumptions to compare the six methods: that PET should be higher than AET for annual values and that there is a linear, stationary relationship between AET and PET. The authors found that PT gave the highest correlation coefficients, and HS gave the lowest. Additionally, annual PET decreased from south to north, from the coast to the Piedmont. Lu et al., (2005) suggested the use of PT, Tc and Hamon because they require fewer parameters and show good correlations with AET.

Lingling et al., (2013) state that there are two hydrological groups for estimating ET: on the water, water surface evaporation, soil evaporation and vegetable transpiration, and on land, basin AET measured using the soil moisture extraction function. In these hydrological models, AET depends on the amount of available water in soil. When soil moisture is high, ET depends on the climate conditions, but water evaporation is high; conversely, when the soil moisture is low, ET decreases.

$$AET = PET\left(\frac{SMT}{SMC}\right) \tag{50}$$

SMT = actual soil moisture; SMC = field capacity soil moisture.

The ratio between AET and PET depends on the availability of water, soil type and LAI (Dyck, 1985; Mintz and Walker, 1993; Lingling et al., 2013).

There are different methods for measuring PET. Gardelin and Lindstrom, (1997) and Lingling et al., (2013) reported that the temperature-corrected Penman equation is reliable, but PT is better because it considers soil heat flux and the negative potential ET in cold weather. Lingling et al., (2013) highlighted that both energy-based and temperature-based methods of estimating PET are more reliable than FAO56PM (Figure 5.1). In particular, Kannan et al., (2007) and Lingling et al., (2013) suggest HG for modeling runoff because FAO56PM requires many data which are frequently difficult to obtain; Oudin et al., (2005) and Oudin et al., (2006) report that inaccurate parameters can make FAO56PM less reliable.

Douglas et al., (2009) recommended estimating PET rather than AET. PET showed an upper limit to ET water losses as a function of available energy, vapor pressure and vegetation type, and the results were more robust than those from using AET; additionally, data layers were available. Yoder et al., (2005) found that for the southeastern United States, FAO56PM gave the best results, and Tc gave good results while being simpler. Moreover, Lu et al., (2005) stated that among PT, Tc and Hamon, PT is better for estimating PET if radiation data are available. Douglas et al., (2009) compared these three models to value which one best represents PET. FAO56PM is the most complex model because of all the weather data, while Tc is the simplest.

The following is the formula for Tc (radiation method for regions with humidity over 50%):

$$\lambda\rho_w ET_o = 0.369\frac{T_{avg}}{T_{avg}+15}\left(2.06R_s+50\right) \tag{51}$$

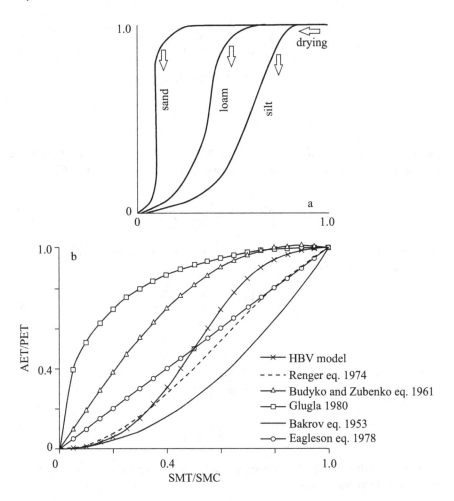

FIGURE 5.1 Nonlinear relationship between AET and PET.

Source: Modified from Lingling et al. (2013).

ET_0 = potential evapotranspiration [mm/day]; λ = latent heat of vaporization [2.451 MJ/kg]; ρ_w = water density [kg/m³]; R_s = daily solar radiation [W/m²]; T_{avg} = mean daily temperature [°C].

PT:

$$\lambda \rho_w ET_o = \alpha \frac{\Delta}{\Delta + \gamma}\left(R_n - G\right) \tag{52}$$

Δ = slope of the saturation vapor pressure–temperature curve; γ = psychrometric constant, R_n = net radiation [W/m²]; G = soil/canopy heat flux [W/m²]:

$$\Delta = \frac{4098e_s}{(273.3 + T_{min})^2} \tag{53}$$

$$\lambda = 2.501 - 0.0002631T_{avg} \tag{54}$$

$$\gamma = \frac{c_p P}{\varepsilon\lambda}10^{-3} = 0.0016286\frac{P}{\lambda} \tag{55}$$

e_s = saturated vapor pressure [kPa]; c_p = specific heat of moist air [=1.013 kJ/kg°C]; P = atmospheric pressure [=101.3 kPa]; T_{min} = minimum daily temperature [°C].

$$e_s = 0.6108 \; exp\left(\frac{17.27T_{min}}{273.3 + T_{min}}\right) \tag{56}$$

The Penman-Monteith (PM) equation for vegetated surfaces considering plant-specific resistance is given as

$$\lambda\rho_w ET_0 = \frac{\Delta(R_n - G) + \rho_a c_p D / r_a}{\Delta + \gamma\left(1 + \dfrac{r_a}{r_s}\right)} \tag{57}$$

D = vapor pressure deficit of the air [kPa]; ρ_a = air density [kg/m³]; r_s = bulk surface resistance [s/m]; r_a = aerodynamic resistance [s/m].

$$\rho_a = 3.486\frac{P}{275 + T_{avg}} \tag{58}$$

P = (constant) 101.3 kPa; T_{avg} = average daily temperature [°C]; $D = e_s - e$; e = observed daily vapor pressure.

$$r_a = \frac{\ln\left[\dfrac{z_u - d}{z_{om}}\right]ln\left[\dfrac{z_e - d}{z_{ov}}\right]}{k^2 u} \tag{59}$$

u = wind speed [m/s]; z_u = height at which wind speed is measured, z_e = height of the vapor pressure/relative humidity instruments; d = displacement height (approximated as 0.67 h_c), h_c = average vegetation height; z_{om} = roughness height for momentum; z_{ov} = roughness height for water vapor (approximated as 0.1 z_{om}); k = von Karman's constant (0.41).

According to Brutsaert and Chen, (1995), PET and daily evapotranspiration (DET) may be compared if soil moisture is not a limiting factor and soil moisture for water-limited PET conditions is not well defined. This transition greatly depends on the soil type, the vegetation and depth where the soil moisture is measured. The authors observed that PM underestimated DET for forests and suggested using at-site surface resistance. PM shows better values for marsh and open water in the same way as Tc because both models are energy based. Based on the daily results, Tc is better when $\beta \leq 1$, and the results from the other two models were poor. Tc and PT overestimated DET, whereas PM underestimated DET; PT and Tc show PET higher than DET.

Drooger and Allen, (2002) highlighted the importance of accurate weather data, in particular for FAO56PM. They compared HG with FAO56PM to determine whether the simpler equation could give good data with inaccurate weather data. Allen et al., (1998) had suggested a different way to determine solar radiation and humidity for FAO56PM but not for wind speed or air vapor pressure. High humidity caused HG to overestimate data, whereas high wind speed caused the equation to underestimate data (Allen et al., 1998; Temesgen et al., 1999; Drooger and Allen, 2002). Hargreaves developed the following equation:

$$ET_o = 0.408*0.0023RA\left(T_{avg}+17.8\right)^{0.5} \tag{60}$$

Droogers and Allen, (2002) stated that HG shows low values in dry regions and high values in wet regions.

Allen et al., (1994) improved HG (Equation (37)) with equation MG1:

$$ET_o = 0.408*0.0030RA\left(T_{avg}+20\right)TD^{0.4} \tag{61}$$

The results are less than 3%, so the authors recommend HG over MG1.

Droogers and Allen, (2002) attempted to improve the equation using the IWMI[1] Climate Atlas data grids (MH2). ET_0 showed higher values in humid areas than FAO56PM. A second attempt based on humidity gave better results; the point was that humidity is not often estimated (Table 5.1 and Figure 5.2).

For monthly values,

$$ET_0 = 0.013 \times 0.408RA(T_{avg} + 17.0)(TD - 0.0123P)^{0.76} \tag{62}$$

P = precipitation [mm]. This equation gave better estimates than FAO56PM, and it is useful because weather data are not always available.

TABLE 5.1

Comparison between Original HG, Original HG with Fitted Parameters and Modified HG

	R^2	RMSD	ET_0	a (mm/d)	b	c	d
Hargreaves	0.895	0.81	2.86	0.0023	17.8		
Hargreaves fitted	0.895	0.79	3.00	0.0025	16.8		
Modified Hargreaves	0.927	0.67	2.96	0.0013	17.0	−0.0123	0.76

Note: Parameters a, b, c and d are the offset multipliers that were used in the HG and modified HG equations.

Source: Modified from Droogers and Allen (2002).

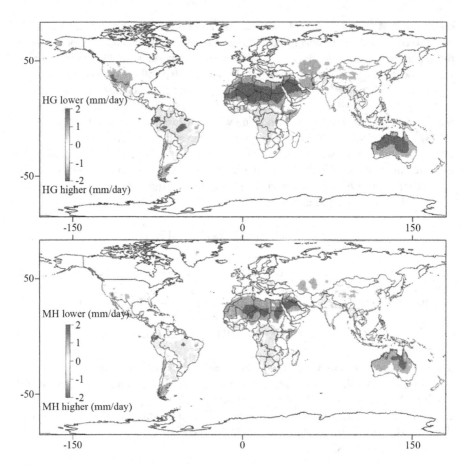

FIGURE 5.2   Differences between annual ET_0 with FAO56PM, HG and modified HG.

Source: Modified from Droogers and Allen (2002).

Based on their findings, Droogers and Allen, (2002) suggest using modified HG (MH) to measure ET_0 using the simpler equation if the weather data are not necessarily accurate, in particular maximum and minimum temperature and precipitation.

Rácz et al., (2013) suggested determining the most suitable method for measuring ET_0 according to the application, noting that the accuracy of each method depends on climate (Federer et al., 1996). In humid climates, researchers suggest using FAO56PM; Jensen et al., 1990; Sumner and Jacobs, 2005; Yoder et al., 2005; Zhou, 2011; Rácz et al., 2013; McMahon et al., 2012 state that the Shuttleworth–Wallace equation (SW) is also effective in humid climates.

$$ET_0 = \frac{C_c ET_c + C_s ET_s}{\lambda} \tag{63}$$

C_c = weighting coefficient for canopy; C_s = weighting coefficient for soil; ET_c = transpiration [mm/day]; ET_s = evaporation [mm/day].

Conversely, Lu et al., (2005) and Adeboye et al., (2009) found PT to be more reliable than the other radiation and temperature methods, although Yates and Strzepek, (1994) and Tabari et al., (2014) found that temperature and radiation methods gave higher values. In arid and semiarid climates, the radiation methods show low values (Er-Raki et al., 2010). With locally calibrated equations, radiation methods perform better than the temperatures methods. Rácz et al., (2013) analized locally calibrated 10 models and compared them, the models were pan coefficient, temperature, radiation and mass transfer based or combined these. The authors chose the models considering the humidity and other climate variables.

In particular, pan-evaporation methods are simple because E_{pan} is estimated, and then the pan coefficient is calculated by a correction factor, so few data inputs are necessary. Even though temperature- and radiation-based methods are simple, using them to estimate empirical coefficients might be difficult. It is suggested to use these methods with monthly or annual values. Mass transfer estimates require the temperature and humidity, but the equations are quite simple. Combination methods are based on the aerodynamic and energy-balance theories. FAO56PM and SW are considered the robust models, but they need a huge amount of data. In particular, SW, a variation of FAO56PM, is useful for local calibration and parameterization but air data, soil and vegetation are fundamental.

NOTE

1. International Water Management Institute.

REFERENCES

Adeboye, O.B., Osunbitan, J.A., Adekalu, K.O., and Okunade, D.A. Evaluation of FAO-56 penman-monteith and temperature based models in estimating reference evapotranspiration using complete and limited data, application to Nigeria. Agric. Engi. Int.: CIGR J., 11, (2009): 1–25.

Allen, R.G., Pereira, L.S., Howell, T.A., and Jensen, M.E. Evapotranspiration information reporting: I. Factors governing measurement accuracy. Agric. Water Manag., 98, (2011): 899–920.

Allen, R.G., Pereira, L.S., Raes, D., and Smith, M. Crop evapotranspiration: Guidelines for computing crop water requirements. In: United Nations FAO, Irrigation and Drainage Paper 56. FAO, Rome, Italy. (1998).

Allen, R.G., Smith, M., Pereira, L.S. An update for the definition of reference evapotranspiration. ICID Bull., 43, (1994): 1–34.

Allen, R.G., Wright, J.L., Pruitt, W.O., Pereira, L.S., and Jensen, M.E. Water requirements. In: Hoffman, G.J., Evans, R.G., Jensen, M.E., Martin, D.L., and Elliot, R.L. (eds.). Design and Operation of Farm Irrigation Systems, 2nd ed. ASABE, St. Joseph, MI, USA, (2007a): 208–288.

Brutsaert, W., and Chen, D.Y. Diagnostics of land surface spatial variability and water vapor flux. J. Geophys. Res., Atmospheres, 100, no D12, (1995): 25595–25606.

Brutsaert, W., and Sugita, M. Application of self-preservation in the diurnal evolution of the surface energy budget to determine daily evaporation. J. Geophys. Res., 97, (1992): 377–382.

Douglas, E.M., Jacobs, J.M., Sumner, D.M., and Ram, L.R. A comparison of models for estimating potential evapotranspiration for Florida land cover types. J. Hydrol., 373, (2009): 366–376.

Droogers, P., and Allen, R.G. Estimating reference evapotranspiration under inaccurate data conditions. Irrig. Drain. Syst., 16, (2002): 33–45.

Dyck, S. Overview on the present status of the concepts of water balance models. IAHS Publ., 148, (1985): 3–19.

Er-Raki, S., Chehbouni, A., Khabba, S., Simonneaux, V., Jarlan, L., Ouldbba, A., Rodriguez, J.C., and Allen, R. Assessment of reference evapotranspiration methods in semi-arid regions: Can weather forecast data be used as alternate of ground meteorological parameters? J. Arid Environ., 74, (2010): 1587–1596.

Federer, C.A., Vörösmarty, C., and Fekete, B. Intercomparison of methods for calculating potential evaporation in regional and global water balance models. Water Resour. Res., 32, (1996): 2315–2321.

Gardelin, M., and Lindstrom, G. Priestley-Taylor evapotranspiration in HBV-simulations. Nord. Hydrol., 28, no 4/5, (1997): 233–246.

Grismer, M.E., Orang, M., Snyder, R., and Matyac, R. Pan evaporation to reference evapotranspiration conversion methods. J. Irrig. Drain. Eng., 128, no 3, (2002): 180–184.

Hamon, W.R. (1963). Estimating potential evapotranspiration. Trans. Am. Soc. Civil Eng., 128, no 1: 324–338.

Hargreaves, G.H., and Samani, Z.A. Reference crop evapotranspiration from temperature. Appl. Eng. Agric., 1, no 2, (1985): 96–99.

Jensen, M.E., Burman, R.D., and Allen, R.G. Evapotranspiration and irrigation water requirements. In: ASCE Manuals and Reports on Engineering Practice. No. 70. (1990).

Kannan, N., White, S.M., Worrall, F., and Whelan, M.J. Sensitivity analysis and identification of the best evapotranspiration and runoff options for hydrological modeling in SWAT-2000. J. Hydrol., 332, no 3/4, (2007): 456–466.

Lingling, Z., Xia, J., Xu, C.-Y., Wang, Z., and Sobkowiak, L. Evapotranspiration estimation methods in hydrological models. J. Geogr. Sci., 23, no 2, (2013): 359–369.

Lu, J., Sun, G., McNulty, S.G., and Amatya, D.M. A comparison of six potential evapotranspiration methods for regional use in the south-eastern United States. J. Am. Water Resour. Assoc., 41, no 3, (2005): 621–633.

Makkink, G.F. Testing the Penman formula by means of lysimeters. J. Inst. Water Eng. Sci., 11, (1957): 277–288.

McMahon, T.A., Peel, M.C., Lowe, L., Srikanthan, R., and Mcvicar, T.R. Estimating actual, potential, reference crop and pan evaporation using standard meteorological data: A pragmatic synthesis. Hydrol. Earth Syst. Sci. Discuss., 9, (2012): 11829–11910.

Mintz, Y., and Walker, G. Global fields of soil moisture and land surface evapotranspiration derived from observed precipitation and surface air temperature. J. Appl. Meteorol., 32, no 8, (1993): 1305–1334.

Oudin, L., Hervieu, F., Michel, C., Perrin, C., Andréassian, V., Anctil, F., and Loumagne, C. Which potential evapotranspiration input for a lumped rainfall-runoff model? (Part 2): Towards a simple and efficient potential evapotranspiration sensitivity analysis and identification of the best evapotranspiration and runoff options for rainfall-runoff modeling. J. Hydrol., 303, no 1–4, (2005): 290–306.

Oudin, L., Perrin, C., Mathevet, T., Andréassian, V., and Michel, C. Impact of biased and randomly corrupted inputs on the efficiency and the parameters of watershed models. J. Hydrol., 320, no 1–2, (2006): 62–83.

Priestley, C.H.B., and Taylor, R.J. On the assessment of surface heat flux and evaporation using large scale parameters. Mon. Weather Rev., 100, (1972): 81–92.

Rácz, C., Nagy, J., and Csaba, D.A. Comparison of several methods for calculation of reference evapotranspiration. Acta Silv. Lign. Hung., 9, (2013): 9–24.

Sumner, D.M., and Jacobs, J.M. Utility of Penman—Monteith, Priestley—Taylor, reference evapotranspiration, and pan evaporation methods to estimate pasture evapotranspiration. J. Hydrol., 308, (2005): 81–104.

Tabari, H., Talaee, P.H., Nadoushani, S.M., Willems, P., and Marchetto, A. A survey of temperature and precipitation based aridity indices in Iran. Quater. Int., 345, (2014): 158–166.

Temesgen, B., Allen, R.G., and Jensen, D.T. Adjusting temperature parameters to reflect well-water conditions. J. Irrig. and Drain. Engrg., 125, no 1, (1999): 26–33.

Thornthwaite, C.W. The moisture-factor in climate. Eos, Trans. Am. Geophys. Union, 27, no 1, (1946): 41–48.

Thornthwaite, C.W. An approach toward a rational classification of climate. Geogr. Rev., 38, no 1, (1948): 55–94.

Turc, L. Estimation des besoins en eau d'irrigation, evapotranspiration potentielle, formule climatique simplifiee et mise a jour. Ann. Agron., 12, no 1, (1961): 13–49.

Yates, D., and Strzepek, K. Potential evapotranspiration methods and their impact on the assessment of river basin runoff under climate change. In: International Institute of Applied Systems Analysis Working Papers. (1994).

Yoder, R.E., Odhiambo, L.O., and Wright, W.C. Evaluation of methods for estimating daily reference crop evapotranspiration at a site in the humid Southeast United States. Appl. Eng. Agric., 21, no 2, (2005): 197–202.

Zhou, M. Estimates of evapotranspiration and their implication in the Mekong and Yellow River Basins. Evapotranspiration, (2011): 319–358.

6 Additional ET Models

6.1 PAN COEFFICIENT-BASED MODEL

The Pereira model (1995) (Per) is given as follows:

$$ET_0 = E_{pan}K_1 \qquad (64)$$

$$K_1 = \frac{0.85(\Delta + \gamma)}{\left[\Delta + \gamma(1 + 0.33u_2)\right]} \qquad (65)$$

u_2 = daily mean wind speed at 2 m height [km/day]; γ = psychrometric constant [kPa/°C]; Δ = slope of the vapor pressure curve [kPa/°C].

FAO56 (Allen et al., 1998) is given here:

$$ET_0 = E_{pan}K_2 \qquad (66)$$

$$K_2 = 0.51206 - (0.000321u_2 + 0.122\ln(F) \\ + 0.1434\ln(RH) - 0.000631[\ln(F)]^2\ln(RH) \qquad (67)$$

u_2 = daily mean wind speed at 2 m height [km/day]; F = the fetch distance above the reference surface [m]; RH = daily mean relative humidity [%].

6.2 TEMPERATURE-BASED MODEL

As a reminder, the Blaney–Criddle model (1950; Doorenbos and Pruitt, 1977a; Burman and Pochop, 1994) is given as follows:

$$ET_0 = a_1 + b_1[p(0.46T + 8.13)] \qquad (8)$$

$$a_1 = 0.0043RH_{min} - (n/N) - 1.41 \qquad (68)$$

$$b_1 = 0.82 - 0.0041RH_{min} + 1.07(n/N) + 0.066u_{2d} \\ - 0.006RH_{min}(n/N) - 0.0006RH_{min}u_{2d} \qquad (69)$$

a_1, b_1 = parameters for the equation; RHmin = daily minimum of relative humidity [%]; (n/N) = relative sunshine duration; u_{2d} = mean wind speed of daylight hours at 2 m height [m/s]; RH = daily mean relative humidity [%]; T = daily mean temperature at 2 m height [°C].

The Szász method (1973) is given as follows:

$$ET_0 = 0.00536(T + 21)^2(1 - RH)^{2/3}f(u) \qquad (70)$$

$$f(u) = 0.0519u_2 + 0.905 \qquad (71)$$

 DOI: 10.1201/9781003467229-6

T = daily mean temperature at 2 m height [°C]; RH = daily mean relative humidity [%]; u_2 = daily mean wind speed at 2 m height [km/day].

6.3   RADIATION-BASED MODEL

Makk-FAO24 (1957; Doorenbos and Pruitt, 1977b) is given as follows:

$$ET_0 = a_2 + b_2 \left(\frac{\Delta}{\Delta + \gamma} \right) \frac{R_g}{\lambda} \tag{72}$$

$$a_2 = -0.3 \tag{73}$$

$$b_2 = c_0 + c_1 RH + c_2 u_{2d} + c_3 RH u_{2d} + c_4 RH^2 + c_5 u_{2d} \tag{74}$$

a2, b2 = parameters for equation; Rg = global radiation [cal/m²day]; Δ = slope of the vapor pressure curve [kPa/°C]; γ = psychrometric constant [kPa/°C]; λ = latent heat of vaporization [cal/m²day]; c_0, c_1, c_2, c_3, c_4, c_5 = coefficients for equation; RH = daily mean relative humidity [%]; u_{2d} = mean wind speed of daylight hours at 2 m height [m/s].

The Priestley–Taylor model (1972; McNaughton and Jarvis, 1983) is

$$ET_0 = \frac{\alpha \dfrac{\Delta}{\Delta + \gamma} (R_n - G)}{\lambda} \tag{75}$$

$$\alpha = 1 + \frac{\gamma}{\Delta + \gamma} \cdot \frac{r_c}{r_a} \tag{76}$$

α = Priestley–Taylor coefficient; Δ = slope of the vapor pressure curve [kPa/°C]; γ = psychrometric constant [kPa/°C]; Rn = net radiation [MJ/m²day]; G = soil heat flux [MJ/m²day]; λ = latent heat of vaporization [MJ/kg]; r_a = aerodynamic resistance [s/m]; r_c = canopy resistance [s/m].

Temeepattanapongsa and Thepprasit, (2015) estimated ET in Thailand by comparing FAO56PM, PT, JH and HG with ET_0 estimated from pan evaporation to determine a reliable alternative for estimating ET_0 according to the climate. The authors conducted the study using 30-year monthly averaged climate data (1966–2011).

The Jensen and Haise (1963) method is presented as follows:

$$ET_0 = K_T (T + T_x) R_S \tag{77}$$

R_S = solar or shortwave radiation [mm/d]; T = mean air temperature [°C].

$$K_T = \cfrac{1}{38 - \dfrac{2 Elev}{305} + \dfrac{36.5}{\left(e^0_{T_{max}} - e^0_{T_{min}} \right)}} \tag{78}$$

$$T_x = 2.5 + 1.4\,(e^0_{T_{max}} - e^0_{T_{min}}) + \frac{Elev}{550} \tag{79}$$

e^0_{Tmax} = saturation vapor pressure at the maximum air temperature; e^0_{Tmin} = saturation vapor pressure at the minimum air temperature; Elev = altitude of the station above mean sea level [m].

Data from FAO56PM showed an average ET of 3.78 mm/d; values from the other equations ranged from 1.59 to 8.75 mm/d (Temeepattanapongsa and Thepprasit, 2015). All the equations except for that using pan evaporation gave higher values than FAO56PM. Comparing data from PT with data from PE, the data were less linear in distribution, maybe due to the constant K_p used in Thailand, although this constant is pan-specific and depends on pan characteristics. PT gave the highest r (correlation coefficient) and the lowest average RE (relative absolute error) and could be improved by adjusting the empirical coefficient. Therefore, PT is the best alternative to FAO56PM for estimating ET_0 in case of missing or limited input data.

6.4   MASS TRANSFER-BASED MODEL

WMO-1966 (WMO66) is given as

$$ET_0 = (0.1298 + 0.0934u_2)(e_s - e_a) \tag{80}$$

u_2 = daily mean wind speed at 2 m height [km/day]; e_s = saturation vapor pressure [hPa]; e_a = actual vapor pressure [hPa].

The Mahringer model (1970) (Mah) is defined as

$$ET_0 = 0.1572\sqrt{(3.6u_2)(e_s - e_a)} \tag{81}$$

u_2 = daily mean wind speed at 2 m height [km/day]; e_s = saturation vapor pressure [hPa]; e_a = actual vapor pressure [hPa].

6.5   COMBINATION METHODS

As a reminder, FAO56PM (F56PM, Allen et al., 1998) is given as

$$ET_0 = \frac{0.408\Delta(R_n - G) + \gamma \frac{900}{T+273}u_2(e_s - e_a)}{\Delta + \gamma(1 + 0.34u_2)} \tag{12}$$

Δ = slope of the vapor pressure curve [kPa/°C]; Rn = water equivalent of net radiation [mm/day]; G = water equivalent of soil heat flux [mm/day]; γ = psychrometric constant [kPa/°C]; u_2 = daily mean wind speed at 2 m height [km/day]; e_s = saturation vapor pressure [kPa]; e_a = actual vapor pressure [kPa]; T = daily mean temperature at 2 m height [°C].

The Shuttleworth–Wallace model (1985) (SW and S&W#2) is given as follows:

$$ET_0 = \frac{C_c ET_c + C_s ET_s}{\lambda}$$ (63)

C_c = weighting coefficient for canopy; C_s = weighting coefficient for soil; ET_c = transpiration [mm/day]; ET_s = evaporation [mm/day]; λ = latent heat of vaporization [MJ/kg].

Rácz et al., (2013) estimated ET for growing season and monthly values. They observed that ET_0 differed according to the model and the year. In particular, the differences were high with methods based on temperature and radiation (see Table 6.1).

The findings showed a strong correlation between Pan and pan-based models due to the use of correction. Conversely, the mass transfer and radiation-based models showed low correlations, possibly because of the diverse atmospheric parameters. The two SW variants showed good agreement. Rácz et al., (2013) also analyzed which models gave the lowest RMSEs, and Per, Szász and PT showed low errors; BC, Szász (temperature), Mak and PT (radiation) showed highly stable responses to changes in both transpiration and humidity. In particular, the curve is inverted because humidity depends on T; the two parameters show a correlation.

Models like Szász, SW, Mak and FAO56PM showed few changes with humidity differences, but mass transfer-based models showed completely different behavior with high humidity. Mak, Szász and PT showed better stability in measuring global radiation sensitivity. Observing the wind speed sensitivity, when the wind speed was below 0.5–2.8 m/s, the models showed similar trends, but Szász, PT, BC, Mak and Per seemed not to be affected by wind speed, whereas mass transfer-based models were affected. Rácz et al., (2013) also observed that the seasonal ET_0 sums were more sensitive to the daily value differences, and these differences can be important during the growing season. Unfortunately, the mass transfer-based models also showed low ET_0 with high standard deviations.

Rácz et al., (2013) observed that Szász, Mak, PT and FAO56PM remained stable and showed little variation when atmospheric parameters changed according to the

TABLE 6.1
Sums of ET_0 during the Growing Seasons of the Examined Years

Year	Pan	Per	FAO56PM	BC	Száz	Makk	PT	WMO66	Mah	SW	F56PM	S&W#2
2005	649	425	517	914	692	861	663	481	582	864	925	**828**
2006	751	493	597	951	708	893	674	520	628	895	948	**853**
2007	1015	644	771	1126	797	1046	746	755	885	1113	1148	**1065**
2008	803	512	628	971	708	923	710	557	657	922	979	**883**
2009	1091	690	830	1153	789	1041	737	799	937	1164	1183	**1117**
2010	751	477	595	877	648	844	683	446	527	799	868	**763**

Note: ET_0 = reference evapotranspiration (mm).

Source: Modified from Rácz et al. (2013).

correlations. In contrast, WMO66, Mar, SW, Per and Pan showed high variations according to the parameter changes. Per and FAO56PM agreed with the pan evaporation data, and SW was very similar to FAO56PM. Considering systematic errors, Mak and SW were close to the pan evaporation, whereas S&W#2, BC and Mak models were close to FAO56PM.

REFERENCES

Allen, R.G., Pereira, L.S., Raes, D., and Smith, M. Crop evapotranspiration: Guidelines for computing crop water requirements. In: United Nations FAO, Irrigation and Drainage Paper 56. FAO, Rome, Italy. (1998).

Blaney, H.F., and Criddle, W.D. Determining water requirements in irrigated areas from climatological and irrigation data. In: Soil Conservation Service Technical Paper 96, US. Department of Agriculture, Washington, USA. (1950).

Burman, R., and Pochop, L.O. Evaporation, Evapotranspiration and Climatic Data. Developments in Atmospheric Science, 22. Elsevier, Amsterdam, (1994).

Doorenbos, J., and Pruitt, W.O. Crop Water Requirements. Fao Irrigation And Drainage. Paper No.24. (Rev.) FAO, Rome, (1977a).

Doorenbos, J., and Pruitt, W.O. Guidelines for Predicting Crop Water Requirements, 2nd ed. FAO, UN, Irrigation and Drainage. Paper No.24. FAO, Rome, (1977b).

Jensen, M.E., and Haise, H.R. Estimating evapotranspiration from solar radiation. J. Irrig. Drain., 89, no 4, (1963): 15–41.

Mahringer, W. Verdunstungsstudien am Neusiedler see. Theor. Appl. Climatol., 18, no 1, (1970): 1–20.

Makkink, G.F. Testing the Penman formula by means of lysimeters. J. Inst. Water Eng. Sci., 11, (1957): 277–288.

Mcnaughton, K.G.; Jarvis, P.G. Predicting effects of vegetation changes on transpiration and evaporation. Water deficits and plant growth, 7, (1983).

Pereira, A.R., Villanova, N., Pereira, A.S., and Baebieri, V.A. A model for the class-A pan coefficient. Agric. Water Manag., 76, (1995): 75–82.

Priestley, C.H.B., and Taylor, R.J. On the assessment of surface heat flux and evaporation using large scale parameters. Mon. Weather Rev., 100, (1972): 81–92.

Rácz, C., Nagy, J., and Csaba, D.A. Comparison of several methods for calculation of reference evapotranspiration. Acta Silv. Lign. Hung., 9, (2013): 9–24.

Shuttleworth, W. J., Gurney, R. J., Hsu, A. Y., & Ormsby, J. P. (1989). FIFE: The variation in energy partition at surface flux sites. IAHS Publ., 186(6), 523–534.

Shuttleworth, W.J., and Wallace, J.S. Evaporation from sparse canopy: An energy combination theory. Q. J. Meteorol. Soc., 111, (1985): 839–855.

Szász, G. A potenciális párolgás meghatározásának új módszere [New method for calculating potential evapotranspiration]. Hidrológiai Közlöny, (1973): 435–442.

Temeepattanapongsa, S., and Thepprasit, C. Comparison and recalibration of equations for estimating reference crop evapotranspiration in Thailand. Agric. Nat. Resour., 49, no 5, (2015): 772–784.

7 Clothesline Effect

Allen et al., (2011) measured ET in a single row of trees with short vegetation, in a strip of cattails along a stream, in an area with short vegetation and a lysimeter placed and without vegetation. Kc was about 1.6–1.8 in midseason for cattails and bulrushes with grass pasture and alfalfa as the vegetation reference, indicating a strong clothesline effect.

This effect is common with ET measured using a lysimeter because of the high vegetation and leaf area in the lysimeter area; this increases aerodynamic and radiative transfer to the canopy and in turn ET. This can be caused when there is more vegetation inside the lysimeter area than there is outside the area. The highest vegetation receives the evaporative energy along the height and laterally with intense exchange of vapor and heat, altering lysimeter measurements. This effect is highest if the lysimeter is smallest (Allen et al., 2011).

Kool et al., (2014) suggested partitioning water based on water use efficiency or water productivity, but others consider T the most important component (Agam et al., 2012; Van Halsema and Vincent, 2012). With full canopy cover, ET is affected by transpiration: biomass correlates with ET. On the contrary, in less vegetated areas, evaporation is the main component of ET. Evaporation is more prominent when water availability and atmospheric demand are high and it can be calculated by removing water by water balancing, by latent heat and/or by measuring water vapor fluxes. Hillel, (1998) identified that water balance is an ET estimation if ET is large and that errors in water balance lead to large errors in ET.

The surface energy balance and the water balance are calculated with the following equation from Chapter 3:

$$R_n - \lambda ET - H - G = 0 \qquad (1)$$

Rn = net radiation [W/m^2]; H = sensible heat [W/m^2]; G = soil heat flux [W/m^2].

Researchers have also considered models with surface energy balance including thermal conversion for the CO_2 fixation, energy advection into the canopy air layer and the rate of storage per unit area in the layer. The energy balance is led by the incoming solar radiation, mostly absorbed from the surface (Brutsaert, 1982). Kool et al., (2014) found that the surface boundary layer was a turbulent region and that the vapor flux's vertical distribution is constant. The authors calculated the equation for the transport of a carrier gas like water vapor:

$$ET = \rho w'q' \qquad (4)$$

ET = [kg/m^2 s]; ρ = air density [kg/m^3]; w' = mean covariance between vertical wind speed and specific humidity q' [m/s]; q' = specific humidity [kg/kg]. This equation is not valid under stable atmospheric conditions (Brutsaert, 1982).

DOI: 10.1201/9781003467229-7

According to Kustas and Agam, (2013), there are many models for measuring E but none that are robust for estimating E under canopy. Kool et al., (2014) identified one model that was based not on a meteorological model but on energy balance, building on work by Ben-Asher et al., (1983). This model is based on comparing the daily minimum and maximum soil surface temperatures of saturated and dry surfaces. Kerridge et al., (2013) confirmed that this model gives reliable E values on days with light wind in a drip-irrigated vineyard.

REFERENCES

Agam, N., Evett, S.R., Tolk, J.A., Kustas, W.P., Colaizzi, P.D., Alfieri, J.G., Mckee, L.G., Copeland, K.S., Howell, T.A., and Chavez, J.L. Evaporative loss from irrigated inter rows in a highly advective semi-arid agricultural area. Adv. Water Res., 50, (2012): 20–30.

Allen, R.G., Pereira, L.S., Howell, T.A., and Jensen, M.E. Evapotranspiration information reporting: I. Factors governing measurement accuracy. Agric. Water Manag., 98, (2011): 899–920.

Ben-Asher, J., Matthias, A.D., and Warrick, A.W. Assessment of evaporation from bare soil by infrared thermometry. Soil Sci. Soc. Am. J., 47, (1983): 185–191.

Brutsaert, W. Evaporation into the Atmosphere: Theory, History and Applications. D. Reidel Publishing, Dordrecht, The Netherlands, (1982).

Hillel, D. Environmental Soil Physics. Academic Press, London, San Diego, CA, (1998).

Kerridge, B.L., Hornbuckle, J.W., Christen, E.W., and Faulkner, R.D. Using soil surface temperature to assess soil evaporation in a drip irrigated vineyard. Agric. Water Manag., 116, (2013): 128–141.

Kool, D., Agama, N., Lazarovitcha, N., Heitmanc, J.L., Sauerd, T.J., and Ben-Gal, A. A review of approaches for evapotranspiration partitioning. Agr. Forest Meteorol., 184, (2014): 56–70.

Kustas, W.P., and Agam, N. Soil evaporation. In: Wang, Y.Q. (ed.). Encyclopedia of Natural Resources. Taylor & Francis, New York, USA, (2013).

Van Halsema, G.E., and Vincent, L. Efficiency and productivity terms for water management: A matter of contextual relativism versus general absolutism. Agric. Water Manag., 108, (2012): 9–15.

8 Estimating ET in Semiarid and Arid Regions

Given the increasing pressure on water resources worldwide, identifying reliable equations for estimating ET_0 in semiarid and arid regions will be critical for increasing the efficiency of agricultural water use. Microirrigation (MI) is an efficient system for delivering water that is increasing in use in developing countries (DehghaniSanij et al., 2004). DehghaniSanij et al., (2004) compared six models with data from lysimeters in Karaj (Iran) with an area in Japan (Tottori) with a temperate climate; the aim was to find the most reliable equation for arid and semiarid climates by measuring the differences between the results from the six equations for the two different climate areas. Generally, ET_0 models are calibrated on long-term weather data from temperate climates, not semiarid and arid areas, for which there is limited information. The ET_0 models considered were PE, FAO56PM, Wright–Penman (WP), BC, radiation balance and HG. Daily weather data were used to produce monthly ET_0. Data from lysimeters (1972 and 1973) were collected from Japan and compared with lysimeter data from Iran.

Penman (1948) first developed this estimation equation:

$$ET_0 = \frac{\left(\frac{\Delta}{\Delta+\gamma}\right)(R_n - G) + K_w \left(\frac{\gamma}{\Delta+y}\right)(a_w + b_w u_2)(e_s - e_a)}{\lambda} \qquad (82)$$

$K_w{}^1$ = constant (6.43); a_w and $b_w{}^2$ = wind function coefficient; u_2 = wind speed at 2 m height[m/s]; λ = latent heat of vaporization [MJ/kg]; $R_n{}^3$ = net radiation at the crop surface [MJ/m² per day]; G^4 = soil heat flux [MJ/m₂ per day]; $(e_s - e_a)^5$ = vapor pressure deficit [kPa]; Δ^6 = slope of vapor pressure curve [kPa/°C]; γ^7 = psychometric constant [0.054 kPa/°C]. WP equation has defined by Wright (1996) developing two equations for the two parameters a_w and b_w in semi-arid climate.

Makkink's model (1957) was modified by Doorenbos and Pruit (1977) (radiation balance):

$$ET_0 = c[(0.25 + 0.50\frac{n}{N})R_a][\frac{\Delta}{(\Delta+\gamma)}] \qquad (83)$$

c = adjustment factor, c depends on mean humidity and daytime wind conditions. The daily values are determined using Allen and Pruitt's (1991) five-parameter model with RH_{mean}, u_2, and Ra (extraterrestrial radiation) [m²/day]. Makkink (1957) measured R_a daily using the solar constant and solar declination data; Δ = slope of vapor pressure curve [kPa/°C]; γ = psychometric constant [0.054 kPa/°C].

DOI: 10.1201/9781003467229-8

Hargreaves and Samani (1985):

$$ET_0 = 0.0023(T_{max} - T_{min})^{0.5}(T_{mean} + 17.8)R_a \qquad (37)$$

Data from the lysimeter showed higher ET from January to July and then a decrease; like a Gaussian curve, it is highest in April. Comparing lysimeter data with results from the model showed that the models produced sometimes low and sometimes high values that were mostly low during October and November, suggesting an association with lower air temperatures. DehghaniSanij et al., (2004) found, however, that other climate parameters can also produce low values. Based on RMSE, FAO56PM gave the most reliable data over the two years; the order was FAO56PM, HG, Wright–Penman (WP), BC, PE and radiation balance.

Based on the mean bias error (MBE), PE, BC and radiation balance gave higher values, whereas HG produced low values, and FAO56PM and WP produced both low and high values. The T-test results showed that FAO56PM performed better than the other models, suggesting that FAO56PM is the best model for measuring ET_0 in semiarid environments. With the lysimeter data from Tottori (Japan) area, all models underestimated ET_0; from low to high ET, the order was PE, radiation balance or HG, WP, FAO56PM and BC. PE was most suitable based on RMSE and MBE, and the t-test findings confirmed the reliability of PE followed by FAO56PM.

$$RMSE = \sqrt{\frac{\sum_{i=1}^{n} d_i^2}{n}} \qquad (84)$$

$$MBE = \frac{\sum_{i=1}^{n} d_i}{n} \qquad (85)$$

$$t = \sqrt{\frac{(n-1)MBE^2}{RMSE^2 - MBE^2}} \qquad (86)$$

d_i = difference between ith predicted and ith measured values; n = number of data pairs i.

It is evident that for humid temperate environments, PE is the most suitable. The work made clear that determining suitable ET_0 models requires weather data from long return periods.

Using lysimeter data from Ankara (Turkey), Benli et al., (2010) calculated RMSE and mean absolute error (MAE) for six equations: FAO56PM, P, FAO24 Radiation (FAO24RD), HG, BC and Class Pan A. Lysimeter data from arid and semiarid environments confirm that FAO56PM is better than other equations (DehghaniSanij et al., 2004; Berengena and Gavilán, 2005; Benli et al., 2006; López-Urrea et al., 2006; Benli et al., 2010), whereas data from ET_0 equations with less intensive data for arid and semiarid environments show less reliable results.

Researchers measured ET using lysimeters and the following equation:

$$ET_a = (S_e - S_b) + d_n + P_e - D_r \qquad (87)$$

ET_a = evapotranspiration [mm]; S_e = water in the root zone at the end of the period [mm]; S_b = water in the root zone at the beginning of the period [mm]; d_n = net irrigation [mm]; P_e = effective precipitation [mm]; D_r = drainage below the root zone [mm].

Lack of humidity data is also a challenge, but actual vapor pressure can be measured by assuming that air temperature is close to T_{min} in the morning and that the air is saturated with water vapor (RH = 100%). Therefore, the temperature at dew point T_{dew} is calculated as

$$e_a = e^0(T_{dew}) = 0.611 \exp\left[\frac{17.27 T_{dew}}{T_{dew} + 237.3}\right] \qquad (88)$$

At low temperatures in arid areas, the air is not always saturated, so to obtain more accurate T_{dew}, T_{min} can be decreased by 2–3 °C (Allen et al., 1998; Benli et al., 2010). FAO56PM very accurately estimates ET_0, particularly when all of the daily data are considered and when the wind speed is closer to the real value. ET_0 is higher when wind speed is 2 m/s, but FAO56PM performed better with wind speed adjusted to 1.2 m/s. HG overestimated ET under high humidity, and solar radiation measured with this equation was lower than the observed radiation data. Both equations show good agreement with lysimeter data. FAO24RD estimated higher ET from the lysimeter than BC, confirming results from López-Urrea et al., (2006), whereas PT showed low values. Class A pan showed low values when ET was high, suggesting that it is more reliable to use climate data to calculate ET_0.

As suggested from FAO, additional ET models need to be validated using FAO56PM, and Mohawesh, (2011) tested the accuracy of eight models against it: PE, Makk, PT, FAO24RD, HG, modified HG 1 (MH1) and modified HG 2 (MH2) (see findings in Table 8.1). Mohawes, (2011) identified that K_c depends on many factors like crop type and growth, canopy height and density but that most FAO56PM data are weather data, and it is not always simple to obtain precise (or sometimes any) data. In these climates, Kc ≈ 1.0–1.1 is reliable because aerodynamic exchange and leaf area for tall vegetation and the alfalfa reference are similar. With the grass reference, Kc exceeded 1.2–1.4 due to the aerodynamic and surface conductance and regional advection.

PE is as follows:

$$ET_0 = \frac{\left(\dfrac{\Delta}{\Delta + \gamma}\right)(R_n - G) + K_w\left[\gamma(\Delta + \gamma)\right]\left(a_w + b_w u_2\right)\left(e_s - e_a\right)}{\lambda} \qquad (89)$$

K_w = (6.43) constant; a_w and b_w = wind function coefficients (empirical constants that were usually computed for regional requirements; values are 1.0 and 0.536, respectively); u_2 = wind speed [m/s], λ = latent heat of vaporization [MJ/kg]. Other notations have the same meaning and units as in FAO56PM.

HG:

$$ET_o = 0.408 * 0.0023\left(T_{mean} + 17.8\right)\left(T_{mean} - T_{min}\right)^{0.5} R_a \qquad (60)$$

MH1:

$$ET_o = 0.408 * 0.0030 \left(T_{mean} + 20\right)\left(T_{mean} - T_{min}\right)^{0.4} R_a \qquad (61)$$

MH2:

$$ET_o = 0.408 * 0.0025 \left(T_{mean} + 16.8\right)\left(T_{mean} - T_{min}\right)^{0.5} R_a \qquad (90)$$

ET_0 = reference evapotranspiration [mm/day]; T_{mean} = mean air temperature [°C]; T_{max} = daily maximum temperature [°C]; T_{min} = daily minimum temperature [°C]; R_a = extraterrestrial radiation [mm/day] from tables (Hargreaves, 1994); constant 0.408 = to convert the radiation to equivalents [mm]; 0.0023 and 17.8 = by fitting measured ET_0 values of the equation (Hargreaves et al., 1985).

FAO24RD:

$$ET_o = a + b\left[\frac{\Delta}{\Delta + \gamma} R_s\right] \qquad (91)$$

a = −0.3 [mm/day]; b = calculated using a regression equation function of RH_{mean} and average day time wind speed.

PT:

$$ET_o = \alpha \frac{\Delta}{\Delta + \gamma}\left(R_n - G\right) \qquad (52)$$

α = 1.26 empirically determined and is a dimensionless correction; other notations have the same meaning and units as in FAO56PM.

Makk is suggested for wet surface areas:

$$ET_o = 0.61\left(\frac{\Delta}{\Delta + \gamma}\right)\left(\frac{R_s}{\gamma} - 0.12\right) \qquad (48)$$

Rs = total solar radiation [MJ/m$_2$/day]; other notations have the same meaning and units as in FAO56PM.

FAO24P (for estimating potential ET from grass):

$$ET_0 = K_p E_{pan} \qquad (92)$$

$$K_P = 0.108 - 0.028U_2 + 0.0422\ln(FET) + 0.1434(RH_{mean})$$
$$-0.000631[\ln(FET)]^2\ln(RH_{mean}) \qquad (93)$$

K_P = pan coefficient; E_{pan} = pan evaporation [mm/day]; U_2 = average daily wind speed at 2 m [m/s], it must be between 1 and 8 m/s; FET = fetch distance of the green crop [m]; RH_{mean} = mean daily relative humidity [%] and must be between 30% and 84%; the fetch distance must be between 1 and 1000 m.

Mohawesh, (2011) estimated the models by calculating statistical parameters. It is observed a huge variation among the models based on the linear regression coefficient of determination (r^2); the data from ET_0 were closer to the FAO56PM results, but unfortunately, r^2 was very poor in arid areas. Analyzing all the data, Mohawesh, (2011) determined that ET_0 has good linear relationships and, also observed high MBEs, RMSEs and MAEs, indicating high bias. PE was less effective for country-wide data, and MH1 and MH2 were more suitable for semiarid than arid regions. HG overestimated ET_0 in humid regions and underestimated it in dry regions, which highlights the need for local calibration for these two equations.

Finally, Mohawesh, (2011) observed that water could be wasted if sap flow is high in arid and semiarid regions; and, irrigation is suggested to be based on ET_0 and Kc values and, adjusting crop water assessments (leaf water potential measurement, plant stem diameter, sap flow data). Farzanpour et al., (2019) confirmed that Tc was suitable for cold and humid climates and arid climates, whereas HS is more accurate for warm and humid and semiarid climates. Accuracy was measured with MAE for calibrated models and relative MAE (RMAE) for non-calibrated models as in Landeras et al., (2008):

$$MAE = \frac{\sum_{i=1}^{N} ET_{i0-ET_{iM}}}{N} \tag{94}$$

$$R\text{-}MAE = 1 - \frac{MAE_{calibrated-model}}{MAE_{non-calibrated-model}} \tag{95}$$

The results confirmed that the temperature- and radiation-based models are more accurate than the mass transfer model, which the authors explained was based on including the temperature and wind speed. Farzanpour et al., (2019) observed also differences in the data according to the season: estimates were more accurate during the cold time than hot time. This could have been due to the ET_0 magnitudes during the seasons, so the results are different if compared with the other models.

Moeletsi et al., (2013) compared HS (Hargreaves-Samani) and Thornthwaite with FAO56PM in South Africa, a semiarid climate using data from eight stations; the commonly measured weather parameters were precipitation and minimum and maximum temperature, whereas parameters such as wind and RH were rare. ET_0 was calculated for 10-day periods based on temperature data, which are recorded at most stations. The authors calibrated both equations with data for the period 1999 to 2004, validated them with data from 2005 to 2008 and compared the findings from both with the FAO56PM results. For each equation, the authors calculated RMSE, MBE and the relative error (RE) for the validation period.

Calibrated HS is given as follows:

$$ET_{CHS} = 0.0135\ C_{HS}\ (T_{av} + 17.8)(T_{max} - T_{min})^{0.5}R_a 0.408d \tag{96}$$

C_{HS} = calibrated HS coefficient.

Calibrated Thornthwaite is given as

$$\mathrm{ET_{CT}} = \mathrm{C_T}\frac{d}{30}\left(10\frac{T_{av}}{1}\right)^{a} \tag{97}$$

C_T = calibrated Thornthwaite coefficient.

HS underestimates $\mathrm{ET_0}$ compared with FAO56PM, so Moeletsi et al., (2013) based their comparison on monthly r^2. The results are good for the growing season from October to April but very low during winter and from May to September. The 10-day data for HS and FAO56PM are very different for each station and for each month. $\mathrm{ET_0}$ is low when the aridity index is over 0.5 and high when the index is lower than 0.5 (Gladderd rift station). Changing C_{HS} improves the results for HS (Moeletsi et al., 2013).

Thornthwaite equation generally underestimated $\mathrm{ET_0}$ based on the high RMSEs and REs. In particular, the 10-day results were very low when r^2 was less than 0.6; the estimates for October, November and December were the best, but May and June were still low. These results confirm that Thornthwaite underestimates $\mathrm{ET_0}$ during the cold season; the original equation showed very poor values at all the stations. HS produced more reliable r^2 results than Thornthwaite, but HS must be used with care during the winter months (June and July).

Azhar et al., (2014) selected five methods HG, modified HG, FAO-radiation, reduced-set Penman-Monteith and FAO-Penman with Watts and Hancock wind function (f(u)). The authors classified these methods as combination or non-combination and as temperature based, temperature-radiation based or combination. They calibrated the equations through coefficients from linear regression analysis from the FAO56PM base data and the $\mathrm{ET_0}$ data sets. The linear regression was

$$\mathrm{ET_{0Base\_method}} = m \times \mathrm{ET_{0method}} + c_1 \tag{98}$$

m = slope of the regression line; c1 = intercept. If m is close to 1 and c1 to 0, correlations in the data are high.

Azhar et al., (2014) observed that adding f(u) to the model overestimated $\mathrm{ET_0}$, mainly because Watts and Hancock's function is a general parameter, not one that is specific to the area. FAO24RD overestimated about 15% of the estimates in semi-arid environments (Berengena and Gavilán, 2005). MH showed poor performance, possibly owing to the advection correction, whereas reduced-set Penman–Monteith showed good results for all months. Calibrating the method improved the $\mathrm{ET_0}$ estimation, in particular correcting for aridity/humidity in the dew point.

Bezerra et al., (2015) calculated daily ET by comparing the results from a remote-sensing algorithm with Bowen ratio measurements for a cotton field in Ceará State, Brazil. Bezerra et al., (2015) estimated daily ET from the surface energy balance algorithm for land (SEBAL) and simplified surface energy balance (SSEB) algorithms and compared the results with ET from BREB. The SSEB algorithm avoids the use of meteorological data except the average temperature between the hot and cold pixels. ET on a field scale depends on the type of vegetation, type of soil, topography,

TABLE 8.1
Comparing ET$_0$ Estimation Models for Countrywide, Arid and Semiarid Data with FAO56PM

Model and region	MAE	RMSE	MBE	r^2	Slope	Intercept
			(mm/day)			
Countrywide						
PE	−4.78	5.48	4.79	0.83	1.59	2.40
Har	−2.75	4.19	3.10	0.41	0.93	3.05
MH1	−0.05	2.12	1.53	0.45	0.46	2.25
MH2	0.00	2.14	1.52	0.44	0.48	2.11
FAO24R	−2.74	3.53	2.96	0.60	0.95	2.96
PT	−2.56	3.80	2.96	0.44	0.87	3.10
Makk	2.25	3.23	2.29	0.48	0.23	0.87
FAO24P	−2.79	4.61	3.45	0.25	0.71	3.94

Model	MAE	RMSE	MBE	r^2	Slope	Intercept
			(mm/day)			
Semiarid area						
PE	−4.10	4.21	4.12	0.91	1.80	1.24
Har	−2.19	3.21	2.42	0.60	1.20	1.46
MH1	−0.11	1.41	1.06	0.64	0.61	1.48
MH2	−0.01	1.39	1.03	0.64	0.64	1.31
FAO24RD	−2.91	3.50	3.01	0.68	1.19	2.23
PT	−2.78	3.75	2.99	0.53	1.14	2.29
Makk	1.82	2.52	1.89	0.57	0.31	0.65
FAO24P	−2.60	3.84	3.03	0.39	0.96	2.72

Model	MAE	RMSE	MBE	r^2	Slope	Intercept
			(mm/day)			
Arid area						
PE	−5.34	6.04	5.34	0.79	1.47	3.24
Har	−3.19	4.83	3.64	0.32	0.77	4.23
MH1	0.02	2.56	1.90	0.35	0.38	2.77
MH2	0.03	2.58	1.90	0.35	0.34	2.67
FAO24RD	−2.60	3.26	2.92	0.57	0.85	3.28
PT	−2.38	3.84	2.93	0.41	0.76	3.45
Makk	2.63	3.72	2.64	0.45	0.20	0.96
FAO24P	−2.93	5.15	3.81	0.17	0.57	4.85

Note: Slope and intercept are from the regression equation.
Source: Modified from Mohawesh (2011).

meteorological conditions and soil water content (Bouwer et al., 2008; Allen et al., 2011; Bezerra et al., 2013). Moreover, Bowen ratio, eddy covariance and soil–water balance are not suitable for larger areas because of the land surface heterogeneity and water vapor transport processes (Gao et al., 2008; Bezerra et al., 2015). Bezerra et al., (2015) suggested using SEBAL, which monitors ET based on the spectral radiance measured using meteorological and satellite sensor data to measure the energy balance at land surface (Bastiaanssen et al., 1998a, 1998b).

SEBAL is based on the evaporative fraction Λ (ratio between latent heat flux and available energy) and on two assumptions: the soil–heat flux in 24 hr is meaningless compared with the net radiation in 24 hr, and Λ is the same as the average daytime value (Shuttleworth et al., 1989; Farah et al., 2004; Bezerra et al., 2015). SEBAL for ET confirms the accuracy of FAO56PM for alfalfa and allows for calculating sensible heat flux (H), atmospheric stability and aerodynamic resistance through micro-meteorological parameters (Gebremichael et al., 2010; Allen et al., 2007a, 2007b; Bastiaanssen et al., 2000; Allen et al., 2002b; Bezerra et al., 2015).

Meanwhile, it is simple to estimate ET using the SSEB algorithm; the latent heat flux depends on the land surface temperature (LST) (Senay et al., 2007, 2013; Bezerra et al., 2015). ET is estimated from the evapotranspiration fraction (ETf), taken from the average temperature of hot and cold pixels (Senay et al., 2007; Bezerra et al., 2015).

ET is calculated from SEBAL (resolution of energy balance equation; Figure 8.1) as follows:

$$Rn = (1-\alpha) \ R_{S\downarrow} + R_{L\downarrow} - R_{L\uparrow} - (1-\varepsilon_0) R_{L\downarrow} \tag{99}$$

Rn = radiation pixel by pixel; α = albedo; $R_{S\downarrow}$ = incoming solar radiation [W/m^2]; $R_{L\downarrow}$ = incoming longwave radiation [W/m^2]; $R_{L\uparrow}$ = outgoing longwave radiation [W/m^2]; ε_0 = surface thermal emissivity.

Soil heat flux (G) is calculated by

$$G = \left[\frac{LST}{\alpha} \left(0.0038\alpha + 0.0074\alpha^2 \right) \left(1 - 0.98NDVI^4 \right) \right] R_n \tag{100}$$

LST was obtained from the modified Planck equation based on the thermal radiance of TM Landsat 5; NDVI: normalized difference vegetation index (Julien et al., 2011). H is the heat loss from the air due to changes in temperature and to convection and conduction, calculated as

$$H = \rho C_p \delta T / r_{ah} \tag{101}$$

ρ = air density [kg/m^3]; C_p = air specific heat [1004 J/kgK] at constant pressure; r_{ah} = aerodynamic resistance to heat transport [s/m] between two near surfaces, z1 and z2 (0.1 m and 2 m, respectively); δT = near surface temperature difference between z1 and z2 [K]. δT is used because it is difficult to measure LST accurately from a satellite due to atmospheric attenuation or contamination and radiometric calibration of the sensor (Allen et al., 2007a, 2007b; Bezerra et al., 2015). This equation is difficult to solve because two parameters are unknown: δT and r_{ah}. Calculating these two parameters and considering also wind speeds high above the ground (in this study 100 m); the wind speeds are then applied in an iterative scheme based on Monin–Obhukov functions (Allen et al., 2007a, 2007b; Bezerra et al., 2015).

Latent heat flux (LE) is the amount of heat loss from the surface because of ET calculated as follows:

$$LE = Rn - G - H \qquad (102)$$

Daily ET was estimated by Λ, calculated as

$$\Lambda = LE/Rn - G \qquad (103)$$

The instantaneous Λ is similar to the value for 24 hr (Shuttleworth et al., 1989; Farah et al., 2004; Brutsaert et al., 1992; Bezerra et al., 2015).
G for a daily period can be meaningless:

$$ET = \Lambda Rn_{24} \qquad (104)$$

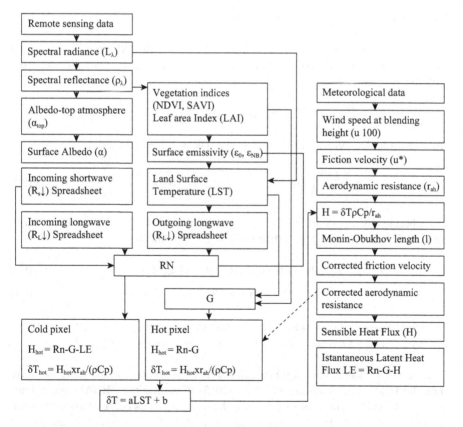

FIGURE 8.1   Flow chart of the SEBAL computation steps.

Source: Modified from Bezerra et al. (2015).

Rn_{24} is the daily net radiation and is calculated based on 24 h integrated meteorological variables:

$$Rn_{24} = (1-\alpha_0)R_{S\downarrow\text{-}24}\text{-}110\tau_{sw24} \ [W/m^2] \tag{105}$$

α_0 = albedo (considered similar to the surface albedo during the morning overpass) (Bezerra et al., 2015); $R_{S\downarrow - 24}$ = 24 h incoming solar radiation; τ_{sw24} = atmospheric transmittance $R_{S\downarrow-24}/R_{0-24}$; R_{0-24} = daily global radiation at the top of the atmosphere.

Daily ET data, estimated by the SSEB algorithm, are calculated using ET_f measured pixel by pixel:

$$ET_f = LST_{HP} - LST/LST_{HP} - LST_{CP} \tag{106}$$

LST_{HP} = average of three selected hot pixels; LST = land surface temperature measured pixel by pixel; LST_{CP} = average of three selected cold selected pixels. Then,

$$ET = ET_f ETr_{(d)} \tag{107}$$

$ETr_{(d)}$ = daily reference evapotranspiration.

Bezerra et al., (2015) collected data to validate the SEBAL from a cotton and castor bean field, in full irrigation conditions and using BREB. They measured the daily ET based on LE:

$$LE = (Rn - G)/(1 + \beta) \tag{108}$$

Rn = net radiation [W/m²]; G = soil heat flux [W/m²]; β = Bowen ratio.

$$G = \left[\frac{LST}{\alpha}\left(0.0038\alpha + 0.0074\alpha^2\right)\left(1 - 0.98NDVI^4\right)\right]Rn \tag{100}$$

LST obtained by the modified Planck equation based on the thermal radiance of TM Landsat 5; NDVI = normalized difference vegetation index.

$$Rn = (1-\alpha)R_s\downarrow + R_L\downarrow - R_L\uparrow - (1-\varepsilon_0)R_L\downarrow \tag{99}$$

$$B = \gamma\delta T/\delta e \tag{109}$$

γ = psychometric constant [kPa/°C]; δT = vertical gradients of air temperature above canopy [°C]; δe = vertical gradients of vapor pressure above canopy [kPa].

Rn and LE (latent heat flux) calculated by SEBAL were much more accurate than the field data, and Bezerra et al., (2015) measured the SEBAL uncertainty based on G and H. The authors proposed that instantaneous energy balance closure is a source of uncertainty in SEBAL and estimated LE as a residual of the energy balance equation. They estimated the residue of the energy balance from field data using the BREB method. The different values are due to the different sensors in the Bowen ratio system and the different values for Rn, G, H and LE (Bastiaanssen et al., 1998a; Bezerra et al., 2015). According to Bezerra et al., (2015), the G uncertainty is not important, and increasing the scale decreases the uncertainty. SEBAL accurately estimated LE and ET, and the authors applied it in different places with the same good performance. Other researchers compared SSEB findings compared with field data from a lysimeter (Gowda et al., 2009; Bezerra et al., 2015). Bezerra et al., (2015) concluded that the SSEB algorithm showed poorer performance than SEBAL (Figure 8.2). Tables 8.2 to 8.4 show the findings for the different SSEB and SEBAL comparisons.

MAE is mean absolute error, MAPE is mean absolute percentage error and RMSE is root mean square error:

$$MAE = N^{-1} \sum_{i=1}^{N} \left| ET(Mod)_i - ET(Obs)_i \right| \tag{110}$$

$$MAPE = N^{-1} \sum_{K=1}^{N} \left| \frac{ET(Mod)_i - ET(Obs)_i}{ET(Obs)_i} \right| \tag{111}$$

$$RMSE = \left[N^{-1} \sum_{i=1}^{N} (ET(Mod)_i - ET(Obs)_i - ET(Obs)_i)^2 \right]^{0.5} \tag{112}$$

TABLE 8.2
Comparison between Surface Energy Fluxes for Satellite Overpass Times and SEBAL Estimates

Date	Crop	Observed (SEBAL) (W/m²)			
		Rn	G	H	LE
09/29/05	Cotton	544 (610)	87 (95)	120 (185)	327 (330)
10/15/05	Cotton	594 (620)	135 (90)	46 (115)	430 (415)
10/15/05	Castor bean	661 (640)	43 (85)	61 (145)	416 (410)
10/31/05	Cotton	589 (650)	109 (68)	72 (180)	391 (402)
10/31/05	Castor bean	631 (660)	81 (100)	172 (190)	401 (370)
11/16/05	Cotton	586 (650)	65 (86)	81 (68)	437 (496)
11/01/08	Cotton	656 (680)	61 (98)	33 (30)	563 (570)
11/17/08	Cotton	680 (690)	115 (104)	62 (72)	492 (515)
12/19/08	Cotton	610 (620)	41 (60)	5 (6)	551 (530)
MBE (W/m²)		41.4	27.0	41.1	19.4
RMSE (W/m²)		40.4	29.6	56.0	25.5
MAPE%		6.9	39.2	62.5	4.4

Source: Modified from Bezerra et al. (2015).

Bezerra et al., (2015) pointed out that the SEBAL and SSEB approaches give lower estimates than the field data because many values are below the one-to-one lines. The SSEB algorithm considers the surface temperature related to soil moisture and atmospheric fluxes, whereas SEBAL is based on the laws of radiation (Iqbal,

TABLE 8.3
Comparing the Fluxes from the Surface Energy on the Field with SSEB Estimates

Date	Crop	ET daily (mm) BREB	ET daily (mm) SEBAL	MAE (mm)	MAPE (%)	RMSE (mm)
09/29/05	Cotton	4.5	4.2	0.33	5.83	0.40
10/15/05	Cotton	5.6	5.2			
10/15/05	Castor bean	5.3	5.3			
10/31/05	Cotton	5.3	5.2			
10/31/05	Castor bean	5.5	5.1			
11/16/05	Cotton	6.2	5.8			
11/01/08	Cotton	6.3	6.2			
11/17/08	Cotton	5.3	5.9			
12/19/08	Cotton	6.9	6.2			

Source: Modified from Bezerra et al. (2015).

TABLE 8.4
Daily ETs in the Cotton Field Using the BREB Method Compared with SEBAL Estimates Based on Landsat 5-TM Images

Date	Crop	Daily ET (mm) BREB	Daily ET (mm) SEBAL	MAE (mm)	MAPE (%)	RMSE (mm)
09/29/05	Cotton	4.5	3.6	0.56	10.20	0.65
10/15/05	Cotton	5.6	6.0			
10/15/05	Castor bean	5.3	5.8			
10/31/05	Cotton	5.3	5.1			
10/31/05	Castor bean	5.5	6.1			
11/16/05	Cotton	6.2	5.0			
11/01/08	Cotton	6.3	6.2			
11/17/08	Cotton	5.3	6.1			
12/19/08	Cotton	6.9	6.6			

Source: Modified from Bezerra et al. (2015).

1983; Liou, 2002; Bezerra et al., 2015), atmospheric corrections, micrometeorology and constancy (Stull, 1988; Brutsaert and Sugita, 1992; Bastiaanssen, et al., 1998a, 1998b; Allen et al., 2007b; Gebremichael et al., 2010). Moreover, SEBAL better estimates ET than SSEB for many different soil coverages.

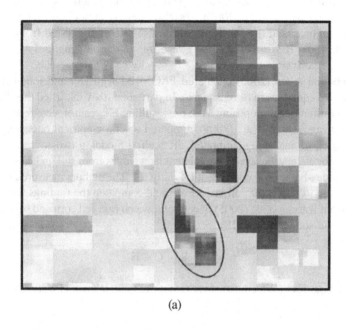

(a)

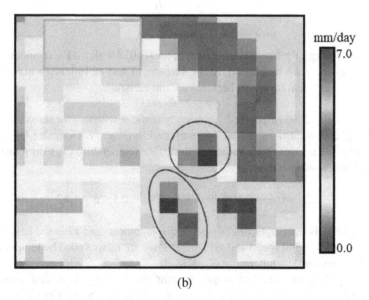

(b)

FIGURE 8.2   Daily ET maps for 19 December 2008 provided by (a) SEBAL and (b) SSEB.
Source: Modified from Bezerra et al. (2015).

Bezerra et al. (2015) highlighted the importance of thermal contamination, which occurs due to differences between the visible spatial resolution and near and thermal infrared channels. Unfortunately, the size of an experimental area could be reduced and could be formed with pixels from an area with different vegetation. If the size of the area is compatible with the spatial resolution of the pixels, the thermal contamination is greater (Allen et al., 2007a, 2007b). SEBAL shows good ET results using the energy balance, and the SSEB algorithm uses Kc as suggested from FAO56PM (Allen et al., 1998; Bezerra et al., 2015), resulting in a promising tool for application in semiarid regions.

Droogers and Allen (2002) compared FAO56PM with HG and found the results from these two equations to be reasonable. In particular, they modified the HG method including a rainfall term proposing MH; the resulting ET_0 values were reliable for arid regions. Gonzalez-Dugo et al., (2009) compared a semi-empirical one source energy balance model, an internally calibrated TR scaling model, a two-source energy balance model for estimating sensible and latent heat fluxes from soil and canopy and the vegetation index-basal-crop coefficient for daily ET. The authors measured differences in net radiation to reflect the soil heat fluxes and compared the findings with a dataset collected from Kutstas et al., (2005) in central Iowa on rain-fed corn and soybean crops.

For the one-source model (1S),

$$LE = R_n - G - H \tag{102}$$

LE = latent heat flux [W/m²]; Rn = net radiation, G = [W/m²]; H = sensible heat flux [W/m²].

$$H = \rho C_p \frac{dT}{r_{AH}} \tag{113}$$

ρ = air density [kg/m³]; Cp = specific heat of air [1005 J/kgK]; dT = temperature gradient [K] between two heights above the surface, z1 and z2 [m]; r_{AH} = aerodynamic resistance to turbulent transport between z1 and z2 [s/m].

Chavez et al., (2005) estimated r_{AH} (aerodynamic resistance) for stable and unstable atmospheric conditions by the Monin–Obukhov theory. Gonzalez-Dugo et al., (2009) proposed a function for radiometric surface temperature (T_R), air temperature (T_A), LAI and wind speed (u):

$$To = 0.534T_R + 0.39T_A + 0.224L - 0.192u + 1.67 \tag{114}$$

The variable z1 is the height of zero-plane displacement, and z2 is the height above the surface where wind speed and air temperature are measured. The temperature at z1 is the aerodynamic temperature (T_o, K).

The z1 height and the roughness lengths for sensible heat and momentum transfer were obtained from SMACEX data (1S-Emp). The 1S-Emp model needs local calibration based on micrometeorological observations. Bastiaansen et al., (1998a) proposed a single-source model, which doesn't need local calibration with micrometeorological data. They approximated the temperature from the linear relationship with the surface temperature:

$$dT = a + bT_R \tag{115}$$

a, b = empirical parameters.

These two parameters were determined by calibrating hot and cold pixels within the satellite (Bastiaansen et al., 1998a). According to Allen et al., (2007b), hot pixels are bare and dry soil; cold pixels are a well-watered crop (ASCE-EWRI, 2005).

Gonzalez-Dougo et al., (2009) studied a two-source (2S) model modified from Norman et al., (1995), Kustas and Norman, (1999) and Li et al., (2005). According to this model, the radiometric surface temperature is the combination of the soil (Ts) and canopy (Tc) temperatures (Figure 8.3):

$$T_R = [f_c T_c^4 + (1 - f_c) T_s^4]^{\frac{1}{4}} \tag{116}$$

fc = fraction of ground covered by the canopy.

They formulated the energy balance equation for the canopy–soil system, canopy and soil layer source:

$$R_n = H + LE + G \tag{117}$$

$$R_{nC} = H_c + LE_c \tag{118}$$

$$R_{nS} = H_s + LE_s + G \tag{119}$$

In the first of these equations, each term is partitioned:

$$R_n = R_{nC} + R_{nS} \tag{120}$$

$$H = H_c + H_s \tag{121}$$

$$LE = LE_c + LE_s \tag{122}$$

Subscripts C and S indicate canopy and soil.

LE_C is determined by PT approximation (Priestley and Taylor, 1972). Under vegetation stress, T_S is too high, causing a non-physical solution for LE_S (Fig 18). LE_S is determined as residual to the energy balance. In this case, soil and canopy fluxes interact:

$$H_c = \rho C_p \frac{T_c - T_{AC}}{r_x} \tag{123}$$

$$H_s = \rho C_p \frac{T_s - T_{AC}}{r_s} \tag{124}$$

$$H = H_c + H_s = \rho C_p \frac{T_{AC} - T_A}{r_A} \tag{125}$$

T_{AC} = air temperature in canopy-air space [K]; r_X = total boundary layer resistance of the canopy of leaves [s/m]; r_S = resistance to heat flow in the boundary layer above the soil surface [s/m]; r_A = aerodynamic resistance to heat transfer [s/m]. The expression for calculating these resistances is in Norman et al., (1995).

Gonzalez-Dougo et al., (2009) highlighted the importance of the daily latent heat rather than the instantaneous value, suggesting three energy balance models. In the first model, the energy budget is preserved during the diurnal cycle (Crago, 1996); the partitioning of energy balance is constant over the day:

$$EF = \frac{LE}{\left(R_n - G\right)} \tag{126}$$

Conversely, according to Gurney and Hsu, (1990) and Brutsaert and Sugita, (1992), EF from morning measures underestimated the daily values because EF increased over the afternoon:

$$EF' = 1.1 \frac{LE}{\left(R_n - G\right)} \tag{127}$$

Allen et al. (2002a, 2002b, 2007a, 2007b) suggested assuming that this fraction is constant during the day (Romero, 2004):

$$ET_{24} = \left(\frac{ET}{ET_o}\right) ET_{024} \tag{128}$$

ET = instantaneous crop evapotranspiration; ET_0 = reference ET; ET_{024} = daily reference ET (according to FAO56PM). Gonzalez-Dugo et al., (2009) also measured daily ET using FAO56PM considering Kc and reference ET according to Doorenbos and Pruitt (1977):

$$ET = (K_{cb}K_s + K_e)ET_0 \tag{129}$$

Ks = reduction in crop transpiration due to soil water deficit; Ke = soil evaporation coefficient (amount of energy available at the soil surface). Doorenboos and Pruitt, (1977) used FAO56PM inputting hourly solar radiation, wind speed, air temperature and relative humidity.

The crop coefficient according to FAO56 (Allen et al., 1998) is calculated as follows:

$$K_e = K_r (K_{cmax} - K_{cb}) \tag{130}$$

K_r = dimensionless evaporation reduction depending on topsoil water depletion; K_{cmax} = maximum Kc following rainfall or irrigation; K_{cb} = basal crop coefficient. K_e cannot exceed the product of $f_{ew} \times K_{max}$, where f_{ew} is the fraction of surface exposed, wet soil surface.

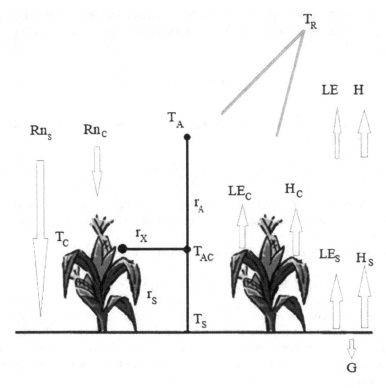

FIGURE 8.3 Schematic illustration of the series resistance network used in the 2S model.

Note: The subscripts c and s indicate canopy and soil, respectively.

Source: Modified from Gonzalez-Dougo et al. (2009).

Vegetation indices (VIs) represent the spectral bands to evaluate vegetation condition, and processes related to the part which is photosynthetically active radiation absorbed by the canopy (fPAR) (Glenn et al., 2008). According to Neale et al. (1989) and Choudhury et al. (1994), K_{cb} and VIs are sensitive to LAI and ground cover fraction (f_c), this confirms the use of spectral measurements:

$$K_{cb} = \frac{K_{cb,mx}}{f_{c,mx}} \left(\frac{SAVI - SAVI_{min}}{SAVI_{max} - SAVI_{min}} \right) \qquad f_c < f_{c,max} \qquad (131)$$

$$K_{cb} = k_{cb,max} \quad f_c \geq f_{c,max} \qquad (132)$$

SAVI = soil-adjusted vegetation index. Max and min refer, respectively, to SAVI for a very large LAI and for bare soil, and $f_{c,max}$ is the f_c at which K_{cb} is maximal ($K_{cb,mx}$).

Calculating K_e and K_s requires measuring soil root zone water balance and determining the occurrence of soil wetting by rainfall. The root zone depth (Zr) was a function of Kcb:

$$Z_r = Z_{rmin} + (Z_{rmax} - Z_{rmin}) \frac{K_{cb}}{K_{cbmax}} \qquad (133)$$

Z_{rmax}= maximum effective root depth; Z_{rmin}= effective root depth during the initial stage of crop growth:

$$\Delta S_w = S_{wf} - S_{wi} = R - ET - D \qquad (134)$$

ΔS_w = change in the root zone water content: S_{wf} = root zone water content (inflow); S_{wi} = root zone water content (outflow); R = infiltrated rainfall; D = deep drainage.

Equation (134) can be expressed as the daily root zone water deficit:

$$RZWD_i = RZWD_{i-1} + ET_i + D_i - R_i \qquad (135)$$

Subscript i = the considered day. $RZWD_i$ = the root zone deficit on day i; $RZWD_{i-1}$ = the root zone deficit on day i–1. If the root zone is watered, RZWD = 0, RZWHC is the water-holding capacity between two extremes: at field level and when the water is at wilting point. The stress coefficient according to the relative root zone water deficit is calculated as

$$K_s = \frac{RZWHC - RZWD_i}{(1-p)RZWHC} \qquad RZDW_i < (1-p)RZWHC \qquad (136)$$

$$K_s = 1 \qquad RZWDI_i \leq (1-p)\ RZWHC \qquad (137)$$

Gonzalez-Dugo et al., (2009) found good agreement among these four models between the estimated and measured values but not sufficient to affect the METRIC performance. These findings aligned with earlier 2S model results obtained by Li et al., (2005) and Gonzalez-Dugo et al., (2006). Moreover, the three energy balance models behave in the same way according to the applied integration method. The 1S model gave higher daily ET, but the METRIC model still gives very good values. The situation improved when the scaling method suggested in Allen et al., (2007b) applied a scaling method that improved the results, but the RMSD increased; in contrast, RMSD and MBE for FAO56PM were close to the energy balance model values (Table 8.5).

The authors determined that the 2S model showed better results than 1S and METRIC (Gonzalez-Dugo et al., 2009). In particular, 1S is not suitable when the meteorological data are not local, contrary to 2S (Kustas and Norman, 1997). Conversely, Norman et al., (2006) and Timmermans et al., (2007) suggested not using this model with internalized calibration according to the selection of cold and hot pixels, and Gonzalez-Dugo et al., (2009) pointed out the importance of atmospheric correction and calibration to obtain accurate LSTs.

TABLE 8.5
Estimated Daily ET Using Three Methods to Scale from Instantaneous to Daily Values

Scaling method	N[a]	RMSD[b] (mm/d)			MBE[c] (mm/d)			r²		
ET$_o$F	27	2S	1S-Emp	METRIC	2S	1S-Emp	METRIC	2S	1S-Emp	METRIC
EF	27	0.74	0.84	0.76	0.38	0.59	−0.08	0.76	0.84	0.76
EF'	27	0.64	0.57	0.92	−0.52	−0.32	−0.08	0.81	0.7	0.76
EF''		0.39		0.58	−0.05	0.27	−0.25	0.81		0.76
FAO-VI	27		0.42			0.01			0.7	

Note: a = number of observations; b = root mean square difference; c = mean bias error; ET$_o$F: reference evapotranspiration fraction; EF: evaporative fraction; EF': 1.1 × EF; FAO-VI: FAO-vegetation index.

Notes: 2S: two-source model, 1S-Emp: one-source empirical model, METRIC: one-source internalized calibration model.

Source: Modified from Gonzalez-Dugo et al. (2009).

Moreover, on clear days, the instantaneous values are suitable as daily values, but the extrapolation must be carefully measured and is not effective under different atmospheric conditions. The 2S model is very effective if rainfall and reference ET data are available at spatial resolution; the results are more reliable than those from FAO56PM for estimating crop ETs under conditions of moisture stress. Figure 8.4 (a–d) shows the findings for each of the four models.

Kamali et al., (2015) compared two different methods of estimating ET$_0$. First they applied HS to weather station data and then interpolated the data; in the second approach, they interpolated the components of the equation and then created an ET$_0$ map of the HS data and commands in GIS. They outlined the reliability of data from distant weather stations, suggesting a suitable interpolation method based on spatial data analysis rather than arithmetic mean or regression, which are both easy but not very reliable methods. The authors also identified two interpolation techniques: deterministic and geostatistical. The former shows a surface from measured data based on the inverse distance weighted or the radial basis functions; the latter measures the correlations between the spatial configurations of the sample points and measured points.

Using geostatistical techniques, a raster map is created from the statistical properties of the measured points; the quality of these prediction maps is determined by the presence of error or uncertainty, inferring the quality of the predictions. Kriging, or variography, is the quantification of spatial correlation, predicting a location without the measure data (Matheron, 1963). Geostatistical methods use different kriging techniques: ordinary, simple, universal, probability, indicator, disjunctive and cokriging. Kamali et al., (2015) used these methods to produce ET$_0$ maps in GIS from spatial data from meteorological stations following two different approaches:

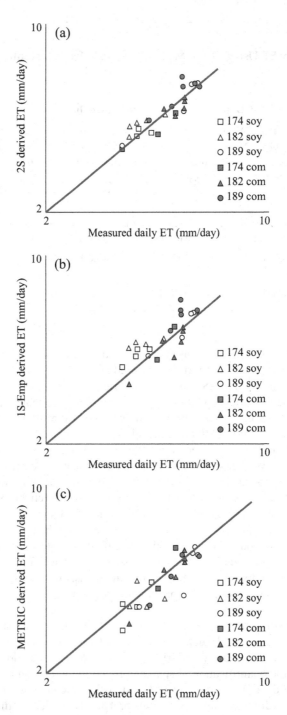

FIGURE 8.4 Daily latent heat estimates produced by the four models compared with the daily values measured by the flux tower network: (a) 2S, (b) 1S-Emp, (c) METRIC, (d) FAO-VI.

Source: Modified from Gonzalez-Dugo et al. (2009).

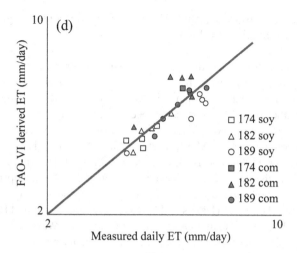

FIGURE 8.4 (Continued)

interpolate and then calculate (IC) and calculate and then interpolate (CI). The outcomes from the two procedures do not differ for a given month and method.

Gradient plus inverse distance squared is used to create these ET_0 maps, and this method shows the most reliable results with the IC procedure (Mardikis et al., 2005). According to Bechini et al., (2000), the spatial and temporal structures are not reliable with IC, and the model produces errors. Kamali et al., (2015) determined that IC is appropriate for nonlinear models paying attention during the interpolation of inputs. In the CI approach, ET_0 is based on climate data, and HS is used; with IC, they interpolated the HS components and prepared the ET_0 maps. Their aim was to estimate these two approaches.

HS (1985) is given as follows:

$$ET_0 = 0.0023 Ra(T + 17.8)TD^{0.5} \tag{37}$$

$$TD = T_{max} - T_{min} \tag{138}$$

$$T = \frac{(T_{max} - T_{min})}{2} \tag{139}$$

Tmax = maximum temperature of the air [°C]; Tmin = minimum temperature of the air [°C]; Ra = extraterrestrial radiation [MJ/m²min].

$$Ra = 37.586 dr(Ws. \sin\theta \sin\delta + \cos\varphi \cos\delta \sin Ws) \tag{140}$$

$$W_s = arccos(-\tan\varphi \tan\delta) \tag{141}$$

$$dr = 1 + 0.033\cos(0.0172J) \tag{142}$$

$$\delta = 0.409\sin(0.0172J - 1.39) \tag{143}$$

$$J = Integer(30.5M - 14.6) \tag{144}$$

dr = inverse relative distance Earth to the Sun; Ws = sunset hour angle; φ = latitude (rad); δ = solar declination; J = Julian day.

Kamali et al., (2015) found that HS was reliable for Iran but suggested applying FAO56PM because data are from the synoptic weather station and fewer pieces of data are needed; data interpolation is also good with HS. The authors also recommended following IC rather than CI interpolation because the former shows good accuracy, whereas the latter is more complex and time consuming, and accuracy depends on the method of computing ET_0 (Mardikis et al., 2005; Ashraf et al., 1997). Conversely, Phillips and Marks, (1996) and Bechini et al., (2000) identified more errors with IC due to errors from the interpolation of each input variable. The appropriateness of the interpolation methods depend on the chosen variables, the spatial configuration of data and number of samples (Isaaks and Srivastava, 1989; Weber and England, 1994; Martinez-Cob, 1996; Caruso and Quarta, 1998; Nalder and Wein, 1998).

Kool et al., (2014) described that evaporation from a drying soil takes place in two or three stages (Lemon, 1956; Ritchie, 1972; Idso et al., 1974). They reported that during stage 1, the atmospheric conditions are fundamental because evaporation depends on the available energy in the upper soil layer and the vapor gradient between soil and atmosphere; in stages 2 and 3, E is a function of soil water content, soil hydraulic characteristics and temperature gradient (Allen et al., 1990; Deol et al., 2012).

Akhavan et al., (2019) measured daily ET for corn using the single ($K_{C\text{-single}}$) and dual ($K_{C\text{-dual}}$) crop coefficients in a semiarid climate in Iran, comparing lysimeter data with daily ET from one combination-based model, one pan evaporation-based model, nine temperature-based models, 10 radiation-based models and seven mass transfer-based models. Climate affects the results from all these models. $K_{C\text{-single}}$ is measured on crop transpiration and soil evaporation as a single datum, whereas $K_{C\text{-dual}}$ separates E from T; the two values separated better evaluated soil wetting due to rain or irrigation. ET_C for the mini-lysimeter was measured using the water balance method:

$$ET_C = P + I - D - R - \Delta S \qquad (145)$$

$$\Delta S = S_t - S_{t-1} \qquad (146)$$

P = rain [mm]; I = irrigation [mm]; D = water loss through drainage from the lysimeter [mm]; R = runoff [mm]; ΔS = change in soil water storage in the lysimeter [mm]; S_t = water in the root zone at the beginning of the period [mm]; S_{t-1} = water in the root zone at the end of the period [mm].

The single-crop coefficient was calculated by FAO56PM: $K_{C\text{ single}} = K_{C\text{ recommended}}$, where $K_{C\text{ recommended}}$ is the K_C from FAO56PM, and it is chosen according to climate conditions. The dual-crop coefficient might show the effects of transpiration from the crop and evaporation from soil:

$$K_{C\text{-dual}} = K_{cb} + K_e \qquad (147)$$

K_{cb} = effect of transpiration from the crop; K_e = effect of evaporation from the soil. Findings from both approaches are presented in Table 8.6.

TABLE 8.6
Mean Crop Coefficients for Each Stage Based on the Single- and Dual-Coefficient Approaches

Crop growth stage	K_{cb}	K_e	$K_{c\text{-single}}$	$K_{c\text{-dual}}$
Initial	0.15	0.33	0.30	0.48
Development	0.70	0.26	0.88	0.99
Mid-season	1.21	0.12	1.35	1.38

Source: Modified from Akhavan et al. (2019).

K_e (soil evaporation coefficient) decreases as the corn grows and ground cover increases. Initially, the evaporation from the soil surface is higher than from the crop.

Comparing ET_C from FAO56PM and the pan evaporation-based models with the daily data from lysimeters, ETc was higher in the initial stage than in the development and mid-season stages. Additionally, the results for daily corn ET_C from the pan evaporation-based model using the single- and dual-crop coefficients are lower than those for the growing season; the pan evaporation-based model and dual-crop coefficient performed better (Table 8.7).

Among the temperature-based models with single- and dual-crop coefficients compared with ET_C obtained by lysimeter, Hargreaves-M3 showed the best results, followed by Hargreaves-M2.

Hargreaves-M3 is as follows:

$$ET_0 = 0.408 \times 0.0013(T_{mean} + 17)(T_{max} - T_{min} - 0.0123p)^{0.76}R_a \qquad (148)$$

T_{mean}= mean temperature [°C]; T_{max}= maximum temperature [°C]; T_{min}= minimum temperature [°C]; R_a= extraterrestrial radiation [MJ/m²day].

Hargreaves-M2 is

$$ET_0 = 0.408 \times 0.0025(T_{mean} + 16.8)(T_{max} - T_{min})^{0.5}R_a \qquad (89)$$

T_{mean} = mean temperature [°C]; T_{max} = maximum temperature [°C]; T_{min} = minimum temperature [°C]; R_a = extraterrestrial radiation [MJ/m²day].

Both Hargreaves-M2 and M1 showed lower corn $ET_{c\text{-single}}$ in the initial development stage, but it was high during the mid-season stage.

Hargreaves-M1 is:

$$ET_0 = 0.408 \times 0.0030(T_{mean} + 20)(T_{max} - T_{min})^{0.4}R_a \qquad (61)$$

T_{mean} = mean temperature [°C]; T_{max} = maximum temperature [°C]; T_{min} = minimum temperature [°C]; R_a = extraterrestrial radiation [MJ/m²day]. Other models, like

TABLE 8.7

Daily Corn ETc Measured with Three Methods (Lysimeter, Pan Evaporation and FAO56PM) Based on Single-Crop and Dual-Crop Coefficients (Iran)

Model	ETc (mm)		RMSE (mm/day)		MMBE (mm/day)		E (mm/day)		d (mm/day)		PE (%)		Rank Score (rank)	
	$K_{c\text{-single}}$	$K_{c\text{-dual}}$	$K_{c\text{-single}}$	$K_{c\text{-dual}}$	$K_{c\text{-single}}$	$K_{c\text{-dual}}$	$K_{c\text{-single}}$	$K_{c\text{-dual}}$	$K_{c\text{-single}}$	$K_{c\text{-dual}}$	$K_{c\text{-single}}$	$K_{c\text{-dual}}$	$K_{c\text{-single}}$	$K_{c\text{-dual}}$
Lysimeter	371	371												
FAO56PM	377	477	2.09	2.48	0.09	1.180	0.07	−2	0.79	0.70	1.61	28.57	0.33 (1)	0.66 (2)
Pan evaporation	255	320	2.51	2.61	−2.1	−0.9	−0.33	−0.07	0.67	0.77	31.26	13.74	0.70 (2)	0.39 (1)

Source: Modified from Akhavan et al. (2019).

Jensen and Haise (1963), Blaney and Criddle (1962) and Baier and Roberston (1965) underestimated the corn ET_c compared with ET_c data from lysimeters.

Jensen and Haise (1963):

$$ET_0 = C_T (T_{mean} - C_x) R_s \tag{149}$$

$$C_x = 22.5 - 0.14(e_{smax} - e_{smin}) - \frac{h}{500} \tag{150}$$

$$C_T = \frac{1}{45 - \left(\dfrac{h}{137}\right) + \left(\dfrac{365}{e_{smax} - e_{smin}}\right)} \tag{151}$$

$$e_{smax} = \exp \frac{19.08 T_{max} + 429.41}{T_{max} + 237.3} \tag{152}$$

$$e_{smin} = \exp \frac{19.08 T_{min} + 429.41}{T_{min} + 237.3} \tag{153}$$

T_{mean} = mean temperature [°C]; R_s = solar radiation [MJ/m^{2}day]; C_t, C_x = empirical coefficient; e_{smax} = maximum saturation vapor pressure [kPa]; e_{smin} = minimum saturation vapor pressure [kPa]; T_{max} = maximum temperature [°C]; T_{min} = minimum temperature [°C].

Blaney and Criddle (1962):

$$ET_0 = 25.4 \frac{(1.8 T_{mean} + 32)}{180} p \tag{154}$$

T_{mean} = mean temperature [°C]; p = constant (0.28).

Baier and Robertson (1965):

$$ET_0 = 0.157 T_{max} + 0.158(T_{max} - T_{min}) + 0.109 R_a + 5.39 \tag{155}$$

T_{max} = maximum temperature [°C]; T_{min} = minimum temperature[°C]; R_a = extraterrestrial radiation [MJ/m^{2}day].

However, these three models all produced low ET_C, and BC and Baier–Roberston (BR) gave low $ET_{C\text{-single}}$ for the growing season. Among the temperature-based models Hargreaves-M3 was the best model for daily corn ET_C, and Hargreaves-M2 was best for $ET_{C\text{-dual}}$; the latter better estimated corn ET than lysimeter data. In contrast, Hargreaves-M2 showed low $K_{C\text{-dual}}$, and Hargreaves-M1 and M3 gave lower values among the temperature-based ET_C models. Hargreaves-M3 gave high corn $ET_{C\text{-dual}}$; Hargreaves-M1, M2 and M3 gave low initial low $ET_{C\text{-dual}}$ but high values during development and mid-season (Akhavan et al., 2019). Hargreaves models perform well in estimating $K_{C\text{-single}}$ and $K_{C\text{-dual}}$ (Chuanyan and Zhongren, 2007; Tabari, 2010; Akhavan et al., 2019) in humid and semiarid conditions, whereas JH, BC and BR give low values for corn ET_{C-dual}. The values are reliable using the dual-crop coefficient and temperature-based models, better than the values from the single-crop coefficient (Akhavan et al., 2019).

Neither the single- nor the dual-crop coefficients performed well with the radiation-based models. Makkink and Heemst, (1967) performed better using K_{C-dual}, whereas the Caprio model (1974) performed well using $K_{C-single}$. The Irmak et al., (2003b), Ritchie (1972) and Makk models performed better using the $K_{C-single}$ model in semiarid climates (Irmak et al., 2003a; Pandey et al., 2016; Trajkovic and Kolakovic, 2009; Akhavan et al., 2019). The radiation-based models using the dual-crop coefficient produced fewer errors than the single-crop coefficient.

Makkink and Heemst (1967):

$$ET_0 = 0.7\frac{\Delta}{\Delta + \gamma}\frac{R_s}{\lambda} \tag{156}$$

R_s = solar radiation [MJ/m^{2}day]; Δ = slope of vapor pressure curve [kPa/$^\circ$C]; γ = psychrometric constant [kPa/$^\circ$C]; λ = latent heat of evaporation [Mj/kg].

Caprio (1974):

$$ET_0 = (0.1092708T + 0.0060706)R_s \tag{157}$$

T = mean temperature [$^\circ$C]; R_s = solar radiation [MJ/m^{2}day].

Ritchie (1972):

$$ET_0 = \alpha[3.87 \times 10^{-3}xR_s(0.6T_{max} + 0.4T_{min} + 29)] \tag{158}$$

$$5 < T_{max} < 35\ \alpha = 1.1 \tag{159}$$

$$T_{max} > 35\ \alpha = 1.1 + 0.05(T_{max} - 35) \tag{160}$$

$$T_{max} < 5\ \alpha = 0.1exp[0.18(T_{max} + 20) \tag{161}$$

$ET_{C-single}$ and ET_{C-dual} calculated with mass transfer-based models are low were the exception of the Romanenko (1961) model, which overestimated ET_C for corn for all of the season using both the single- and dual-crop coefficients (Akhavan et al., 2019).

Meyer (1926), Mahringer (1970) and WMO (1966) showed lower corn $ET_{C-single}$ than did ET_C measured with lysimeters, whereas Albrecht's (1950) model performed well with the dual-crop coefficient.

Romanenko (1961):

$$ET_0 = 4.5\{1 + (\frac{T_{mean}}{25})\}^2(1 - \frac{e_a}{e_s}) \tag{162}$$

T_{mean} = mean temperature [$^\circ$C]; e_a = actual vapor pressure [kPa]; e_s = saturation vapor pressure [kPa].

Meyer (1926):

$$ET_0 = (0.375 + 0.050206U)(e_s - e_a) \tag{163}$$

e_a = actual vapor pressure [kPa]; e_s = saturation vapor pressure [kPa]; U = wind speed [m/s].

Mahringer (1970):

$$ET_0 = 0.15072\sqrt{3.6U}\,(e_s - e_a) \tag{164}$$

U = wind speed [m/s]; e_a = actual vapor pressure [kPa]; e_s = saturation vapor pressure [kPa].

WMO (1966):

$$ET_0 = (0.1298 + 0.0934U)(e_s - e_a) \tag{79}$$

U = wind speed [m/s]; e_a = actual vapor pressure [kPa]; e_s = saturation vapor pressure [kPa].

Albrecht (1950):

$$ET_0 = (0.1005 + 0.297u)\,(e_s - e_a) \tag{165}$$

u = wind speed [m/s]; e_a = actual vapor pressure [kPa]; e_s = saturation vapor pressure [kPa].

According to Akhavan et al., (2019), the single-crop coefficient was more reliable using the temperature- and radiation-based models, and results were worst for mass transfer. Moreover, $K_{c\text{-dual}}$ showed better results than $K_{c\text{-single}}$, and values produced from ET_c using $K_{c\text{-dual}}$ performed better. In arid and semiarid climates, when there are insufficient available data to use FAO56PM, we recommend using simpler equations to obtain reliable data.

NOTES

1. Small and ignored for day period (Gday $\cong$ 0); calculated according to the FAO procedures.
2. Calculated using daily RHmax, RHmin, Tmax, Tmin and Tmean according to the FAO procedure.
3. Calculated evaluated using daily RHmax, RHmin, Tmax, Tmin and Tmean according to the FAO procedure.
4. Calculated according to the FAO procedures.
5. Calculated according to the FAO procedures.
6. Calculated using daily RHmax, RHmin, Tmax, Tmin and Tmean according to the FAO procedure.
7. Calculated evaluated using daily RHmax, RHmin, Tmax, Tmin and Tmean according to the FAO procedure.

REFERENCES

Akhavan, S., Kanani, E., and Dehghanisanij, H. Assessment of different reference evapo-transpiration models to estimate the actual evapotranspiration of corn (Zea mays L.) in a semiarid region (case study, Karaj, Iran). Theor. Appl. Climatol., 137, (2019): 1403–1419.

Albrecht, F. Die Methoden zur Bestimmung Verdunstung der naturlichen Erdoberfläche. Arc. Meteor. Geoph. Bioklimatol. Ser. B, 2, (1950): 1–38.

Allen, R.G., Morse, A., Tasumi, M., Trezza, R., Bastiaanssen, W.G.M., Wright, J.L., and Kramber, W. Evapotranspiration from a satellite-based surface energy balance for the Snake River Plan aquifer in Idaho. In: Proceeding of the USCID/EWRI Conference on Energy, Climate, Environment, and Water. U.S. Committee on Irrigation and Drainage, Denver, CO, USA. (2002a).

Allen, R.G., Pereira, L.S., Howell, T.A., and Jensen, M.E. Evapotranspiration information reporting: I. Factors governing measurement accuracy. Agr. Water Manag., 98, (2011): 899–920.

Allen, R.G., Pereira, L.S., Raes, D., and Smith, M. Crop evapotranspiration: Guidelines for computing crop water requirements. In: United Nations FAO, Irrigation and Drainage Paper 56. FAO, Rome, Italy. (1998).

Allen, R.G., and Pruitt, W.O. FAO-24 reference evapotranspiration factors. J. Irrig. Drain. Eng., 117, no 5, (1991): 758–773.

Allen, R.G., Tasumi, M., and Trezza, R. SEBAL (Surface Energy Balance Algorithms for Land), Advanced Training and User's Manual, Idaho Implementation, Version 1.0, USA. (2002b).

Allen, R.G., Tasumi, M., and Trezza, R. Satellite-based energy balance for Mapping Evapotranspiration with Internalized Calibration (METRIC) model. J. Irrig. Drain. Eng., 133, (2007b): 380–394.

Allen, R.G., Wright, J.L., Pruitt, W.O., Pereira, L.S., and Jensen, M.E. Water requirements. In: Hoffman, G.J., Evans, R.G., Jensen, M.E., Martin, D.L., and Elliot, R.L. (eds.). Design and Operation of Farm Irrigation Systems, 2nd ed. ASABE, St. Joseph, MI, USA, (2007a): 208–288.

Allen, S.J. Measurement and estimation of evaporation from soil under sparse barley crops in Northern Syria. Agric. For. Meteorol., 49, (1990): 291–309.

ASCE-EWRI the ASCE standardized reference evapotranspiration equation. In: ASCE-EWRI Standardization of Reference Evapotranspiration Task Committee Rep., ASCE, Reston, VA, USA. (2005).

Ashraf, M., Loftis, J.C., and Hubbard, K.G. Application of geostatistics to evaluate partial weather station networks. Agr. For. Meteorol., 84, (1997): 255–271.

Azhar, A.H., Masood, M., Nabi, G., and Basharat, M. Performance evaluation of reference evapotranspiration equations under semiarid Pakistani conditions. Arab. J. Sci. Eng., 39, (2014): 5509–5520.

Baier, W., and Robertson, G.W. Estimation of latent evaporation from simple weather obser-vations. Can. J. Plant Sci., 45, no 3, (1965): 276–284.

Bastiaanssen, W.G.M. SEBAL-based sensible and latent heat fluxes in the irrigated Gediz Basin, Turkey. J. Hydrol., 229, (2000): 87–100.

Bastiaanssen, W.G.M., Menenti, M., Feddes, R.A., and Holtslag, A.A.M. A remote sensing Surface Energy Balance Algorithm for Land (SEBAL) 1: Formulation. J. Hydrol., 212–213, (1998a): 198–212.

Bastiaanssen, W.G.M., Pelgrum, H., Wang, J., Ma, Y., Moreno, J.F., Roenrink, G.J., and van der Wal, T. A remote sensing Surface Energy Balance Algorithm for Land (SEBAL) 2., Validation. J. Hydrol., 212–213, (1998b): 213–229.

Bechini, L., Ducco, G., Donatelli, M., and Stein, A. Modelling, interpolation and stochastic simulation in space and time of global solar radiation. Agric. Ecosyst. Environ., 81, (2000): 29–42.

Benli, B., Bruggeman, A., Oweis, T., and Üstün, H. Performance of Penman-Monteith FAO56 in a semiarid highland environment. J. Irrig. Drain. Eng., 136, no 11, (2010): 757–765.

Benli, B., Kodal, S., Ilbeyi, A., and Ustun, H. Determination of evapotranspiration and basal crop coefficient of alfalfa with a weighing lysimeter. Agric. Water Manag., 81, no 3, (2006): 358–370.

Berengena, J., and Gavilán, P. Reference evapotranspiration estimation in a highly advective semiarid environment. J. Irrig. Drain. Eng., 131, no 2, (2005): 147–163.

Bezerra, B.G., da Silva, B.B., dos Santos, C.A.C., and Bezerra, J.R.C. Actual evapotranspiration estimation using remote sensing: Comparison of SEBAL and SSEB approaches. Adv. Remote Sens., 4, (2015): 234–247.

Bezerra, B.G., Santos, C.A.C., Silva, B.B., Perez-Marin, A.M., Bezerra, M.V.C., Bezerra, J.R.C., and Ramana Rao, T.V. Estimation of soil moisture in the root-zone from remote sensing data. Rev. Bras. de Ciênc. do Solo, 37, (2013): 595–603.

Blaney, H.F., and Criddle, W.D. Determining consumptive use and irrigation water requirements (No. 1275). US Department of Agriculture. (1962).

Bouwer, L.M., Biggs, T.W., and Aerts, C.J.H. Estimates of spatial variation in evaporation using satellite-derived surface temperature and a water balance model. Hydrol. Process., 22, (2008): 670–682.

Brutsaert, W., and Sugita, M. Application of self-preservation in the diurnal evolution of the surface energy budget to determine daily evaporation. J. Geophys. Res., 97, (1992): 377–382.

Caprio, J.M. The solar thermal unit concept in problems related to plant development and potential evapotranspiration. In: Lieth, H. (ed.). Phenology and Seasonality Modeling, Ecological Studies, Vol. 8, Springer Verlag, New York, USA, (1974): 353–364.

Caruso, C., and Quarta, F. Interpolation methods comparison. Comput. Math. Appl., 35, no 12, (1998): 109–126.

Chavez, J.L., Neale, C.M.U., Hipps, L.E., Prueger, J.H., and Kustas, W.P. Comparing aircraft-based remotely sensed energy balance fluxes with eddy covariance tower data using heat flux source area functions. J. Hydromteorol., 6, (2005): 923–940.

Choudhury, B.J., Ahmed, N.U., Idso, S.B., Reginato, R.J., and Daughtry, C.S.T. Relations between evaporation coefficients and vegetation indices studied by model simulations. Remote Sens. Environ., 50, (1994): 1–17.

Chuanyan, Z., and Zhongren, N. Estimating water needs of maize (Zea mays L.) using the dual crop coefficient method in the arid region of northwestern China. Afr. J. Agric. Res., 2, no 7, (2007): 325–333.

Crago, R.D. Conservation and variability of the evaporative fraction during the daytime. J. Hydrol., 180, (1996): 173–194.

DehghaniSanij, H., Yamamotoa, T., and Rasiah, V. Assessment of evapotranspiration estimation models for use in semi-arid Agricultural. Water Manag., 64, no 2, (2004): 91–106.

Deol, P., Heitman, J.L., Amoozegar, A., Ren, T., and Horton, R. Quantifying nonisothermal subsurface soil water evaporation. Water Resour. Res., 48, (2012): 1–11.

Doorenbos, J., and Pruitt, W.O. Guidelines for predicting crop water requirements. In: FAO, UN, Irrigation and Drainage Paper No.24., 2nd Ed., FAO, Rome, Italy. (1977).

Droogers, P., and Allen, R.G. Estimating reference evapotranspiration under inaccurate data conditions. Irrig. Drain. Syst., 16, (2002): 33–45.

Farah, H.O., Bastiaanssen, W.G.M., and Feddes, R.A. Evaluation of the temporal variability of the evaporative fraction in a tropical watershed. Int. J. Appl. Earth Obs. Geoinf., 5, (2004): 129–140.

Farzanpour, H., Shiri, J., Sadraddini, A.A., and Trajkovic, S. Global comparison of 20 reference evapotranspiration equations in a semi-arid region of Iran. Hydrol. Res., 50, no 1,
(2019): 282–300.

Gao, Y., Long, D., and Li, Z. Estimation of daily evapotranspiration from remotely sensed
data under complex terrain over the upper Chao River Basin in North China. Int. J.
Remote Sens., 29, (2008): 3295–3315.

Gebremichael, M., Wang, J., and Sammis, T.W. Dependence of remote sensing evapotranspiration algorithm on spatial resolution. Atmos. Res., 96, (2010): 489–495.

Glenn, E., Huete, A., Nagler, P., and Nelson, S. Relationship between remotely sensed
vegetation indices, canopy attributes and plant physiological processes: What vegetation indices can and cannot tell us about the landscape. Sensors, 8, no 4, (2008):
2136–2160.

Gonzalez-Dugo, M.P., Neale, C.M.U., Mateos, L., Kustas, W.P., and Li, F. Comparison of
remote sensing-based energy balance methods for estimating crop evapotranspiration.
In. Rem. Sens. for Agric. Ecosyst. Hidrol.VIII, 6359, (2006): 218–226.

Gonzalez-Dugo, M.P., Neale, C.M.U., Mateos, L., Kustas, W.P., Prueger, J.H., Anderson,
M.C., and Li, F. A comparison of operational remote sensing-based models for estimating crop Evapotranspiration. Agr. For. Meteorol., 149, (2009): 1843–1853.

Gowda, P.H., Senay, G.B., Howell, T.A., and Marek, T.H. Lysimetric evaluation of simplified
surface energy balance approach in the Texas high plains. Appl. Eng. Agric., 25, (2009):
665–669.

Gurney, R.J., and Hsu, A.Y. Relating evaporative fraction to remotely sensed data at FIFE
site. In: Symposium on FIFE: Fist ISLSCP Field Experiment, February 7–9. American
Meteorological Society, Boston, USA, (1990): 112–116.

Hargreaves, G.H. Defining and using reference evapotranspiration. J. Irrig. Drain. Eng., 120,
no 6, (1994): 1132–1139.

Hargreaves, G.L., Hargreaves, G.H., and Riley, J.P. Agricultural benefits for Senegal River
basin. J. Irrig. Drain. Engr., ASCE, 111, no 2, (1985): 113–124.

Hargreaves, G.H., and Samani, Z.A. Reference crop evapotranspiration from temperature.
Appl. Eng. Agric., 1, no 2, (1985): 96–99.

Idso, S.B., Reginato, R.J., Jackson, R.D., Kimball, B.A., and Nakayama, F.S. Three stages of
drying of a field soil. Soil Sci. Soc. Am. J., 38, (1974): 831–837.

Irmak, S., Irmak, A., Allen, R.G., and Jones, J.W. Solar and net radiation based equations to
estimate reference evapotranspiration in humid climates. J. Irrig. Drain. Eng., 129, no 5,
(2003b): 336–347.

Irmak, S., Irmak, A., Jones, J.W., Howell, T.A., Jacobs, J.M., Allen, R.G., and Hoogenboom,
G. Predicting daily net radiation using minimum climatological data. J. Irrig. Drain.
Eng., 129, no 4, (2003a): 256–269.

Iqbal, M. An Introduction to Solar Radiation. Academic Press, Toronto, Canada, (1983).

Isaaks, E.H., and Srivastava, R.M. An Introduction to Applied Geostatistics. Oxford
University Press, NewYork, USA, (1989).

Jensen, D.T., Hargreaves, G.H., Temesgen, B., and Allen, R.G. Computation of ET_0 under non
ideal conditions. J. Irrig. Drain. Eng., 123, (1997): 394–400.

Jensen, M.E., and Haise, H.R. Estimating evapotranspiration from solar radiation. J. Irrig.
Drain., 89, no 4, (1963): 15–41.

Julien, Y., Sobrino, J.A., Mattar, C., Ruescas, A.B., Jiménez-Muñoz, J.C., Sòria, G., Hidalgo,
V., Atitar, M., Franch, B., and Cuenca, J. Temporal analysis of Normalized Difference
Vegetation Index (NDVI) and Land Surface Temperature (LST) parameters to detect
changes in the Iberian Land Cover between 1981 and 2001. Int. J. Remote Sens., 32,
(2011): 2057–2068.

Kamali, M.E., Nazari, R., Faridhosseini, A., Ansari, H., and Eslamian, S. The determination of reference evapotranspiration for spatial distribution mapping using geostatistics. Water Resour. Manage., 29, (2015): 3929–3940.

Kool, D., Agama, N., Lazarovitcha, N., Heitmanc, J.L., Sauerd, T.J., and Ben-Gal, A. A review of approaches for evapotranspiration partitioning. Agr. For. Meteorol., 184, (2014): 56–70.

Kustas, W.P., Hatfield, J., and Prueger, J.H. The Soil Moisture Atmosphere Coupling Experiment (SMACEX): Background, hydrometerological conditions and preliminary findings. J. Hydrometeorol., 6, (2005): 791–804.

Kustas, W.P., and Norman, J.M. A two-source approach for estimating turbulent fluxes using multiple angle thermal infrared observations. Water Resour. Res., 33, (1997): 495–1508.

Kustas, W.P., and Norman, J.M. Evaluation of soil and vegetation heat flux predictions using a simple two-source model with radiometric temperatures for partial canopy cover. Agric. For. Meteorol., 94, no 1, (1999): 13–29.

Landeras, G., Ortiz-Barredo, A., & López, J. J. (2008). Comparison of artificial neural network models and empirical and semi-empirical equations for daily reference evapotranspiration estimation in the Basque Country (Northern Spain). Agricultural water management, 95(5), 553–565.

Lemon, E.R. The potentialities for decreasing soil moisture evaporation loss. Soil Sci. Soc. Am. J., 20, (1956): 120–125.

Li, F., Kustas, W.P., Prueger, J.H., Neale, C.M.U., and Jackson, J.T. Utility of remote sensing based two-source energy balance model under low and high vegetation cover conditions. J. Hydrometerol., 6, (2005): 878–891.

Liou, K.N. An Introduction to Atmospheric Radiation, 2nd ed. Academic Press, San Diego, USA, (2002).

López-Urrea, R., de Santa Olalla, F.M., Fabeiro, C., and Moratalla, A. Testing evapotranspiration equations using lysimeter observations in a semiarid climate. Agric. Water Manag., 85, no 1–2, (2006): 15–26.

Mahringer, W. Verdunstungsstudien am Neusiedler see. Arch. Met. Geoph. Biokl. Ser. B, 18, (1970): 1–20.

Makkink, G.F. Testing the Penman formula by means of lysimeters. J. Inst. Water Eng. Sci., 11, (1957): 277–288.

Makkink, G.F., and Heemst, H.V.Potential evaporation from short grass and water. (1967): 89–96.

Mardikis, M.G., Kalivas, D.P., and Kollias, V.J. Comparison of interpolation methods for the prediction of reference evapotranspiration—An application in Greece. Water Resour. Manag., 19, (2005): 251–278.

Martinez-Cob, A. Multivariate geostatistical analysis of evapotranspiration and precipitation in mountain terrain. J. Hydrol., 174, (1996): 19–35.

Matheron, G. Principles of geostatistics. Econ. Geol., 58, (1963): 1246–1266.

Meyer, A. Über einige Zusammenhänge zwischen Klima und Boden in Europa. Chem. Erde, 2, (1926): 209–347.

Moeletsi, M.E., Walker, S., and Hamandawana, H. Comparison of the Hargreaves and Samani equation and the Thornthwaite equation for estimating dekadal evapotranspiration in the free state province, South Africa. Phys. Chem. Earth, Parts A/B/C., 66, (2013): 4–15.

Mohawesh, O.E. Evaluation of evapotranspiration models for estimating daily reference evapotranspiration in arid and semiarid environments. Plant Soil Environ., 57, no 4, (2011): 145–152.

Nalder, I.A., and Wein, R.W. Spatial interpolation of climatic normal: Test of a new method in the Canadian boreal forest. Agric. For. Meteorol., 92, (1998): 211–225.

Neale, C.M.U., Bausch, W.C., and Heermann, D.F. Development of reflectance based crop coefficients for corn. Trans. ASAE, 32, no 6, (1989): 1891–1899.

Norman, J.M., Anderson, M.C., and Kustas, W.P. Are single-source, remote-sensing surface-flux models too simple? In: D'Urso, G., Osann Fochum, M.A., and Moreno, J. (eds.). *Proceeding of the International Conference on Earth Observation for Vegetation Monitoring and Water Management*, Vol. 852. AIP, (2006): 170–177.

Norman, J.M., Kustas, W.P., and Humes, K.S. A two-source approach for estimating soil and vegetation energy fluxes in observations of directional radiometric surface temperature. Agric. For. Meteorol., 77, (1995): 263–293.

Pandey, P.K., Dabral, P.P., and Pandey, V. Evaluation of reference evapotranspiration methods for the northeastern region of India. J. Soil Water Conserve., 4, no 1, (2016): 56–67.

Penman, H.L. Natural evaporation from open water, bare soil, and grass. Proceeding of Royal Society of London, UK, A193, (1948): 120–145.

Phillips, D.L., and Marks, D. Spatial uncertainty analysis: Propagation of interpolation errors in spatially distributed models. Ecol. Model., 91, (1996): 213–229.

Priestley, C.H.B., and Taylor, R.J. On the assessment of surface heat flux and evaporation using large scale parameters. Mon. Weather Rev., 100, (1972): 81–92.

Ritchie, J.T. Model for predicting evaporation from a row crop with incomplete cover. Water Resour. Res., 8, (1972): 1204–1213.

Romanenko, V.A. Computation of the autumn soil moisture using a universal relationship for a large area. Proc. Ukr. Hydrometeorol. Res., Inst., 3, (1961).

Romero, M.G. Daily evapotranspiration estimation by means of evaporative fraction and reference evapotranspiration fraction. Ph.D. Dissertation, Utah State Univ. Logan, Utah, USA. (2004).

Senay, G.B., Bohms, S., Singh, R.K., Gowda, P.H., Velpuri, N.M., Alemu, H., and Verdin, J.P. Operational evapotranspiration mapping using remote sensing and weather datasets: A new parameterization for the SSEB approach. J. Am. Water Res. Assoc., 49, (2013): 577–591.

Senay, G.B., Budde, M., Verdin, J.P., and Melesse, A. A coupled remote sensing and simplified surface energy balance approach to estimate actual evapotranspiration from irrigated fields. Sensors, 7, (2007): 979–1000.

Shuttleworth, W.J., Gurney, R.J., Hsu, A.Y., and Ormsby, J.P. FIFE: The variation in energy partition at surface flux sites. IAHS Publ., 186, (1989): 67–74.

Stull, R.B. An Introduction to Boundary Layer Meteorology. Kluwer Academic Publishers, Boston, USA, (1988).

Tabari, H. Evaluation of reference crop evapotranspiration equations in various climates. Water Res Manag., 24, (2010): 2311–2337.

Timmermans, W.J., Kustas, W.P., Anderson, M.C., and French, A.N. An intercomparison of the Surface Energy Balance Algorithm for Land (SEBAL) and the Two Sources Energy Balance (TSEB) modeling schemes. Remote Sens. Environ., 108, (2007): 284–369.

Trajkovic, S., and Kolakovic, S. Evaluation of reference evapotranspiration models under humid conditions. Water Resour. Manag., 23, no 14, (2009): 3057–3067.

Weber, D., and England, E. Evaluation and comparison of spatial interpolators II., Math. Geol., 26, (1994): 589–603.

WMO Measurement and estimation of evaporation and evapotranspiration. In: Technical Paper (CIMO-Rep.) No. 83. Genova, Switzerlands. (1966).

Wright, J.L., Derivation of alfalfa and grass reference evapotranspiration in evapotranspiration and irrigation scheduling. In: Camp, C.R., Saddler, E.J., Yoder, R.E. (Eds.), Proceedings of the ASAE International Conference on Evapotranspiration and Irrigation Scheduling, 3–6 November, San Antonio, TX, (1996): 133–140.

9 Estimating ET in Humid Regions

Bogawski and Bednorz, (2014) compared four simple evaporation equations in Poland with FAO56PM: radiation, aerodynamic (mass transfer), temperature and pan evaporation (Xu and Singh, 2002; Mallikarjuna et al., 2014). Twelve equations were derived from the literature, and five methods were developed. Following are the four equations:

$$\text{(FAO56)PM\_ANG: } R_n = R_{ns} - R_{nl} \tag{34}$$

R_{ns} = net shortwave radiation [MJ/m²day]; R_{nl} = net longwave radiation [MJ/m²day].

$$R_{ns} = 0.77 R_s \tag{166}$$

R_s = solar radiation [MJ/m²day].

$$R_s = (a_s + b_s \frac{n}{N}) R_a \tag{32}$$

R_a = extraterrestrial radiation [MJ/m²day] calculated according to Allen et al., (1998); N = maximum possible sunshine duration [hour]; n = actual sunshine duration [hour]; a_s and b_s = regression constants. Allen et al., (1998) suggested two values for the two constants if there are not calibrated values: $a_s = 0.25$ and $b_s = 0.5$. On clear days, $n = N$.

$$R_{sclear} = 0.75 R_a \tag{167}$$

$$R_{nl} = \sigma \left[\frac{T_{max,K}^4 + T_{min,K}^4}{2} \right] \left(0.34 - 0.14\sqrt{e_a} \right) \left(1.35 \frac{\left(a_s + b_s \frac{n}{N} \right) R_a}{0.75 R_a} - 0.35 \right) \tag{168}$$

σ = Stefan–Boltzmann constant [4.903 10⁻⁹ MJ/K⁴ m² day]; $T_{max,K}$ = maximum absolute temperature during a 24 hr period [K = °C+273.16]; $T_{min,K}$ = minimum absolute temperature during a 24 hr period [K = °C+273.16].

PMF_ANG_adj:

$$R_s = (0.094 + 0.549 \frac{n}{N}) R_a \tag{169}$$

DOI: 10.1201/9781003467229-9

R_s is adjusted on data from a linear regression between R_s/R_a (clearness index) and n/N relative sunshine duration (Sabziparvar et al., 2013; Bogawski and Bednorz, 2014):

Priestley–Taylor method:

$$ET_o = \alpha \frac{\Delta}{\Delta + \gamma}(R_n - G)\frac{1}{\lambda} \qquad (45)$$

$\alpha = 1.26$ empirical coefficient latent heat.

This multiple regression model used R_n, T_{max} and T_{min} as independent variables, and FAO56PM is a dependent variable; T_{min} was considered insignificant and eliminated. This model was not good, and Bogawski and Bednorz, (2014) applied Cochrane–Orcutt estimation (CO) (Thejll and Schmit, 2005; Wen, 2009).

CO method:

$$ET_0 = -0.755 + 0.257R_n + 0.0627T_{max} \qquad (170)$$

The authors used these methods when radiation or sunshine duration data were absent.

For aerodynamic-based estimations, the authors applied Dalton's equation for water vapor turbulence transfer from an evaporating surface to the atmosphere (Bogawski & Bednorz, 2014):

$$ET_0 = (0.3648 + 0.07223u)(e_s - e_a) \qquad (171)$$

WMO (1966):

$$ET_0 = (0.1298 + 0.0934u)(e_s - e_a) \qquad (79)$$

Mahringer (1970) (model for Austria):

$$ET_0 = 0.15072 \sqrt{3.6u}\,(e_s - e_a) \qquad (80)$$

PE simplified:

$$ET_0 = 0.35(1+0.24u)\,(e_s - e_a) \qquad (172)$$

Penman_adj:

$$ET_0 = 0.36(1 + 0.14u)(e_s - e_a) \qquad (173)$$

For temperature-based estimation, researchers use temperature or temperature combined with precipitation beginning with Hargreaves:

Hargreaves_FAO (Allen et al., 1998):

$$ET_o = 0.408 * 0.0023\left(T_a + 17.8\right)\left(T_{max} - T_{min}\right)^{0.5} R_a \qquad (60)$$

Hargreaves_globe (Droogers and Allen, 2002):

$$ET_o = 0.408 \times 0.0025 (T_a + 16.8)(T_{max} - T_{min})^{0.5} R_a \tag{89}$$

Hargreaves_precipitation (when temperature and precipitation are available):

$$ET_o = 0.408 \times 0.0013 (T_a + 17.0)((T_{max} - T_{min})$$
$$- 0.0123P)^{0.5} R_a \tag{174}$$

P = precipitation [mm].

Bogawski and Bednorz, (2014) modified HG for the Balkans (Trajkovic and Kolakovic, 2007) as Hargreaves_Balk and Poland Hargreaves _adj:

$$ET_o = 0.408 \times 0.0023 (T_a + 17.8)(T_{max} - T_{min})^{0.424} R_a \tag{175}$$

$$ET_o = 0.408 \times 0.001 (T_a + 17.0)(T_{max} - T_{min})^{0.724} R_a \tag{176}$$

For pan evaporation-based estimation, starting from the FAO model:

$$ET_0 = KET_{pan} \tag{177}$$

K_p = empirical coefficient; ET_{pan} = pan evaporation [mm].

Allen et al., (1998) suggested for Colorado:

$$K_p = 0.87 + 0.119\ln\{FA\}(FET) - 0.0157[ln(86.4u)]^2 - 0.019\ln$$
$$(RH_a) - 0.000053\ln(86.4u)\ln\{FA\}(FET)RH_a \tag{178}$$

FET = distance from the evaporimeter to the surrounding low-growing vegetation [m]; RH_a = daily average air relative humidity [%].

For Poland:

$$K_{p\_adj} = 0.569 + 0.078\ln\{FA\}(FET) - 0.0103[ln(86.4u)]^2 - 0.012\ln$$
$$(RH_a) - 0.000035\ln(86.4u)\ln\{FA\}(FET)RH_a \tag{179}$$

From all of these equations, seven methods produced warm season ET_0 totals (Figure 9.1); the other methods confirmed the FAO56PM data. Radiation-based methods give the most reliable results, followed by mass transfer, pan coefficient and temperature.

For monthly ET, Bogawski and Bednorz, (2014) found that three methods were suitable: FAO56PM-ANG_adj, CO and Hargreaves_adj; the one-month findings with the other methods were over- or underestimates. Dalton gave the best results among the aerodynamic equations, although it overestimated ET_0 for August and

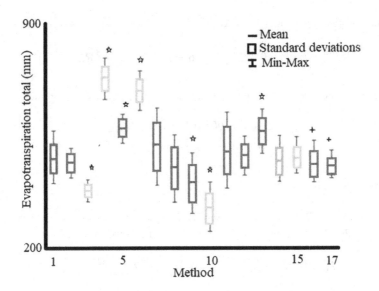

FIGURE 9.1 Evapotranspiration totals for warm season (April–October) in Sulejów (1981–2010) calculated by 17 different methods.

Notes: Values differ from reference PMF56 ET based on Tukey's post hoc test, *p < 0.001; + data for shorter period (May–October; 1999–2010): 1—PMF56, 2—Hargreaves_adj, 3—Hargreaves_precip, 4—Hargreaves_globe, 5—Hargreaves_Balk, 6—Hargreaves_FAO, 7—Penman, 8—Penman_adj, 9—Mahringer, 10—WMO, 11—Dalton, 12—PMF_ANG_adj, 13—PMF_ANG, 14—Priestley-Taylor, 15—CO, 16—Pan coefficient_FAO, 17—Pan coefficient_adj.

Source: Modified from Bogawski and Bednorz (2014).

October compared with FAO56PM. Mah and WMO underestimated four and six months, their results are very poor. The pan coefficient gave reliable results for the summer period, but the results were over- or underestimated in the other months. HG was the least suitable model, but the Hargreaves_adj shows reliable results; the other methods showed results far from the FAO56PM values.

Bogawski and Bednorz, (2014) identified four ET equations that estimated better than FAO56PM in case of missing data: HG, FAO56PM-ANG, Penman_adj and Pan coefficient_adj, although for the last two methods, the improvements were small. Based on the Willmott index, Hargreaves_adj showed fewer errors. The Hargreaves_adj, FAO56PM-ANG_adj and CO results agreed wih FAO56PM; the other methods gave mistakes in one of the months: PT for October; Penman_adj for May and June, Pan coefficient for May. The Penman methods overestimated August, September and October data; Penman_adj and Dalton October data and Pan coefficient for September and October data.

Xu et al., (2006) andYin et al., (2008) previously established that solar radiation is the most influencing factor in ET equations. According to Tabari et al., (2014) radiation-based multiple regression is highly suited to the humid climate in Iran; conversely, Bogawski and Bednorz, (2014) found that HG, FAO56PM-ANG, CO and Hargreaves_adj performed better daily, annually and monthly in humid areas. FAO56PM-ANG showed better daily values than CO. When solar radiation and/or sunshine duration are lacking, Penman_adj showed better daily values. Dalton and PE were suitable without calibration. In short, combining vapor pressure deficit and wind speed allows for reliably describing the evapotranspiration process.

Bogawski and Bednorz, (2014) established that regional calibration is mandatory for using the temperature-based methods in Poland and Central Europe; they found that the accuracy diminished with decreasing data availability. If radiation data are absent and sunshine duration data are present, PMF_ANG_adj is the most suitable for ET_0. If sunshine radiation data are also unavailable, Penman_adj using vapor pressure deficit and wind speed data is suitable in this region. For humid climates, Bogawski and Bednorz, (2014) recommend FAO24RD as a radiation-based method and HS (Hargreaves-Samani) as a temperature-based method.

FAO24 Blaney-Criddle is an effective temperature based-method in the sub-humid and semiarid climates in India. Nikam et al., (2014) compared monthly and seasonal times obtained using HG and Thorntwaite as temperature methods and PT and Tc as radiation methods with FAO56PM. To estimate ET_0, these methods need maximum, minimum and average temperature. On the contrary, Thornthwaite and HG miss data such as RH (minimum, average and maximum), average wind speed, sunshine hours, solar radiation and net radiation. The authors compared these methods using the following regression analysis equation:

$$Y = mX + C \tag{180}$$

Y = estimated monthly ET_0 [mm/day]; X = standard ET_0 from each of the four methods [mm/day]; m = slope of the equation; C = intercept on the regression equation.

The authors calculated RMSE, absolute average deviation (AAD) and absolute relative error (ARE) as follows:

$$AAD = \frac{\sum_{i=1}^{n} ABS(y_i - x_i)}{n} \tag{181}$$

$$ARE = \frac{ABS(y_i - x_i)}{x_i} \tag{182}$$

$$RMSE = \sqrt{\frac{\sum_{i=1}^{n}(y_i - x_i)}{n}} \tag{183}$$

The most suitable method for estimating ET is the one with the lowest AAD and m (slope of the equation) close to 1, a low RMSE and high R^2 (see Table 9.1). Checking for both monthly and seasonal results, Nikam et al., (2014) found that Tc performed better than HG on a monthly basis, although R^2 was better with the latter; estimate errors were better with Tc, however. For seasonal data, the authors analyzed three seasons: rabi from November to March, summer from April to June and Kharif from July to October. PT and HG showed the best estimate errors during rabi; PT was close to FAO56PM. During Kharif, Tc produced the lower estimate errors, and in the summer, HG did so.

TABLE 9.1
Estimated ET$_0$ for the Four Considered Methods

Data available	ET estimation method			
	HS	TH	TC	PT
	Applicability of the method			
T	y	y	n	n
T + RH	y*	y*	y	y
T + RH + Rad/SS	y*	y*	y*	y
T + RH + Rad/SS + W	y*	y*	y*	y*

Note: T = temperature; RH = relative humidity; Rad = radiation; SS = sunshine hours; W = average wind velocity. y = method can be applied using these data; y* = this method does not use some of the available parameters; n = method cannot be applied due to insufficient data.

Source: Modified from Nikam et al. (2014).

In brief, these data indicate that Tc performs better on a monthly scale, and the total Tc values are close to standard annual ET$_0$, although HG did show good results as well (Nikam et al., 2014).

Chowdhury et al., (2017) analyzed seven ET$_0$ equations to determine the most suitable one for humid areas using FAO56PM as the reference; for each method, they calculated SEE as R^2. Three methods were temperature based: BC, Thornthwaite and HS, with BC being based on monthly temperature data estimated from daily temperature data and on monthly percentage of annual daylight hours from the latitude of the place. Thornthwaite and HS require knowing the temperature and location of the place. Chowdhury et al., (2017) used three radiation-based models as well: Tc, PT (Priestly-Taylor) and FAO24RD.

Thornthwaite had the highest values but did not display the same pattern as the other ET$_0$ equations because it is entirely temperature based, in contrast PT showed higher values. The results showed that ET was low on November and December (winter season) and from June to September (monsoon season), because of the low temperatures and low evaporation rates, high relative humidity and low evaporation. Thornthwaite produced the lowest R^2 and highest estimate errors. BC and HS produced highly similar R^2 and estimate errors. Tc produced the closest values to FAO56PM, whereas Thornthwaite's values were the farthest from FAO56PM.

Samaras et al., (2014) analyzed 18 radiation-based equations to measure ET$_0$ in humid, sub-humid and semiarid Mediterranean conditions. The Mediterranean climate shows high humidity and low wind speed, and the PT, Makk and JH groups showed better performance than the Abtew equation. Radiation-based equations performed better than equations based on temperature (Lu et al., 2005; Gebhart et al., 2013; Samaras et al., 2014). The considered area from the authors was in Greece, and there were no solar radiation data, so they calculated ET with the Ångström–Prescot formula:

$$R_s = a_s + b_s \frac{n}{N} R^a \tag{32}$$

R_s = global total solar radiation; R_a = extraterrestrial radiation; n/N = relative sunshine duration; a_s and b_s = constant (0.22 and 0.52).

For Lefkada and Nafpaktos stations:

$$R_s = K_{Rs} \sqrt{(T_{max} - T_{min})} \, R_a \qquad (184)$$

T_{max} = maximum temperature [°C]; T_{min}= minimum temperature [°C]; Rs, Ra [MJ/m²day].

The first radiation-based equation was Christiansen (1968) (Hargreaves and Allen, 2003):

$$ET_{ref} = 0.385 \frac{R_s}{\lambda} \qquad (185)$$

Calibrated Christiansen (Abtew, 1996) (warm, humid climate):

$$ET_{ref} = K \frac{R_s}{\lambda} 0.52 \le K \le 0.54 \qquad (186)$$

JH (Jensen–Haise group 1963) (semiarid to arid):

$$ET_{ref} = (0.025 \, T_{mean} + 0.08) \frac{R_s}{\lambda} \qquad (187)$$

Stephens Stewart (1963, Jensen, 1966) (warm and humid):

$$ET_{ref} = (0.0148 T_{mean} + 0.07) \frac{R_s}{\lambda} \qquad (188)$$

Stephens (1965), (Jensen, 1966) (warm and humid)

$$ET_{ref} = (0.0158 T_{mean} + 0.09) \frac{R_s}{\lambda} \qquad (189)$$

Caprio (1974) (semiarid to arid):

$$ET_{ref} = \frac{25.4}{10^5} (1.8 T_{mean} + 1) R_S \qquad (190)$$

HG (Hargreaves eq.) (1975), (Hargreaves and Allen, 2003):

$$ET_{ref} = (0.0135 T_{mean} + 0.2403) \frac{R_s}{\lambda} \qquad (191)$$

Makk (1957), (Jensen, 1966) (cool and humid):

$$ET_{ref} = 0.61 \frac{\Delta}{\Delta + \gamma} \frac{R_s}{\lambda} - 0.12 \qquad (192)$$

Calibrated Makk (Castañeda and Rao, 2005) (semiarid and arid):

$$ET_{ref} = 0.70 \frac{\Delta}{\Delta + \gamma} \frac{R_s}{\lambda} - 0.12 \qquad (193)$$

De Bruin (1981), (de Bruin and Lablans, 1998) (cool and humid):

$$ET_{ref} = 0.65\frac{\Delta}{\Delta+\gamma}\frac{R_s}{\lambda} \tag{194}$$

PT (Priestly-Taylor eq.) (1972) (humid):

$$ET_{ref} = 1.26\frac{\Delta}{\Delta+\gamma}\frac{R_n-G}{\lambda} \tag{195}$$

Calibrated PT (Abtew, 1996) (warm and humid):

$$ET_{ref} = 1.18\frac{\Delta}{\Delta+\gamma}\frac{R_n-G}{\lambda} \tag{196}$$

Calibrated PT (Berengena and Gavilán, 2005) (semiarid):

$$ET_{ref} = 1.65\frac{\Delta}{\Delta+\gamma}\frac{R_n-G}{\lambda} \tag{197}$$

Modified PT (Xu and Singh, 2000) (humid continental climate):

$$ET_{ref} = 0.98\frac{\Delta}{\Delta+\gamma}\frac{R_n-G}{\lambda}-0.94 \tag{198}$$

TC (Turc eq.) (1961), (Xu et al., 2008) (humid):

$$ET_{ref} = 0.013\frac{T_{mean}}{T_{mean}+15}(R_s+50) \tag{199}$$

Modified TC (Abtew, 1996):

$$ET_{ref} = 0.012\frac{T_{mean}}{T_{mean}+15}(R_s+50) \tag{200}$$

Abtew (1996) (warm and humid):

$$ET_{ref} = \frac{1}{56}\frac{R_s T_{max}}{\lambda} \tag{201}$$

T_{mean} = mean air temperature [°C]; G = soil heat flux density [MJ/m²day]; Δ = slope of saturation vapor pressure curve [KPa/°C]; γ = psychrometric constant [KPa/°C]. Rn and Rs unit measures are MJ/m²day or cal/cm²day for Tc, modified Tc and Caprio. From these 18 equations, Samaras et al., (2014) built a generalized model of seven equations (see Table 9.2).

To validate these equations, Samaras et al., (2014) calculated MAE, RMAE, RMSE and coefficient of efficiency (EF). P_i and O_i are the model-predicted and observed

TABLE 9.2
Radiation-Based Equations and the Final Group Utilized for the Research

Method	Original equation
Christiansen group	Christiansen (1968), Abtew (1996)
Jensen–Haise group	Jensen and Haise (1963), Stephens and Stewart (1963), Stephens (1965), Caprio (1974), Hargreaves (1975)
Makkink group	Makkink (1957), de Bruin (1981), Hansen (1984), Castañeda and Rao (2005)
Priestley–Taylor group	Priestley and Taylor (1972), Abtew (1996), Berengena and Gavilán (2005), Xu and Singh (2000)
Turc	Turc (1961)
Modified Turc	Abtew (1996)
Abtew	Abtew (1996)

Note: a and b were calculated so that the sum of the squares of the residuals would be the smallest.

Source: Modified from Samaras et al. (2014).

(from FAO56PM) variables, respectively; Õ is the observed mean, and N is the number of cases.

$$MAE = N^{-1} \sum_{i=1}^{N} |P_i - O_i| \qquad (202)$$

$$RMAE = (MAE/\bar{O})100 \qquad (203)$$

$$RMSE = [N^{-1} \sum_{i=1}^{N} (P_i - O_i)^2]^{0.5} \qquad (204)$$

$$EF = 1 - \frac{\sum_{i=1}^{N}(P_i - O_i)^2}{\sum_{i=1}^{N}(\bar{O} - O_i)^2} \qquad (205)$$

The authors applied 12-fold cross validation; they divided the data into 12 sets and used the data from 11 for calibrating and to estimate the coefficients. They used the final set for validation based on the differences between the model-predicted and measured values. They also compared their ET_{ref} results from equations with measurements from the lysimeters and with FAO56PM. All the models showed good results; only the Christiansen group showed a high error (Samaras et al., 2014). Christiansen uses Rs, whereas the other models use T_{mean} and T_{max} or Δ. After the calibration, the Makkink equations showed good results.

Comparing the original coefficients from each model with the calibrated coefficients, Samaras et al., (2014) determined that the differences were due to the climate. Climate affected the results from each equation; results were sometimes different in similar climates, and adjusting the general models to the local climatic conditions reduced the errors. Among all the models, Abtew showed good performance under semiarid, sub-humid and humid/moderately windy conditions. When average daily maximum temperature is missing, an alternative was the Jensen–Haise group model, whereas for humid and lightly windy conditions, the Priestley-Taylor group was the most reliable, followed by the JH group. Radiation-based models, in particular models with a temperature parameter, performed very well. Finally, Samaras et al., (2014) recommended Abtew for the Mediterranean climate and the Jensen-Haise, Makkink and Priestley groups are reliable for humid and light windy climate conditions.

REFERENCES

Abtew, W. Evapotranspiration measurements and modeling for three wetland systems in south Florida. J. Am. Water Resour. Assoc., 32, no 3 (1996): 465–473.

Allen, R.G., Pereira, L.S., Raes, D., and Smith, M. Crop evapotranspiration: Guidelines for computing crop water requirements. In: United Nations FAO, Irrigation and Drainage Paper 56. FAO, Rome, Italy. (1998).

Berengena, J., and Gavilán, P. Reference evapotranspiration estimation in a highly advective semiarid environment. J. Irrig. Drain. Eng. 131, no 2, (2005): 147–163.

Bogawski, P., and Bednorz, E. Comparison and validation of selected evapotranspiration models for conditions in Poland (Central Europe). Water Resour. Manage., 28, (2014): 5021–5038.

Caprio, J.M. The solar thermal unit concept in problems related to plant development and potential evapotranspiration. In: Lieth, H. (ed.). Phenology and Seasonality Modeling, Ecological Studies, Vol. 8. Springer Verlag, New York, USA, (1974): 353–364.

Castañeda, L., and Rao, P. Comparison of methods for estimating reference evapotranspiration in southern California. J. Environ. Hydrol., 13, no 14, (2005): 1–10.

Chowdhury, A., Gupta, D., Das, D.P., and Bhowmick, A. Comparison of different evapotranspiration estimation techniques for Mohanpur, Nadia district, West Bengal. International Journal of Computer Engineering Research, 7, no 4, (2017): 33–39.

Christiansen, J.E. Pan evaporation and evapotranspiration from climatic data. J.Irrig. Drain. Div., 94, no 2, (1968): 243–266.

de Bruin, H. The determination of (reference crop) evapotranspiration from routine weather data. In: Proceedings of Technical Meeting 38, Evaporation in Relation to Hydrology, Committee for Hydrological Research TNO, The Hague, Netherlands, 28. (1981): 25–37.

de Bruin, H., and Lablans, W.N. Reference crop evapotranspiration determined with a modified Makkink equation. Hydrol. Process., 12, no 7, (1998): 1053–1062.

Droogers, P., and Allen, R.G. Estimating reference evapotranspiration under inaccurate data conditions. Irrig. Drain. Syst., 16, (2002): 33–45.

Gebhart, S., Radoglou, K., Chalivopoulos, G., and Matzarakis, A. Evaluation of potential evapotranspiration in central Macedonia by EmPEst. In: Helmis, C., and Nastos, P. (eds.). Advances in Meteorology, Climatology and Atmospheric Physics, Vol. 2. Springer Atmospheric Sciencies, Springer, Berlin Heidelberg, (2013): 451–456.

Hansen, S. Estimation of potential and actual evapotranspiration: Paper presented at the nordic hydrological conference (Nyborg, Denmark, August-1984). Hydrol. Res., 15, no 4–5, (1984): 205–212.

Hargreaves, G.H. Moisture availability and crop production. ASCE Trans., 18, no 5, (1975): 980–984.

Hargreaves, G.H., and Allen, R.G. History and evaluation of Hargreaves evapotranspiration equation. J. Irrig. Drain. Eng., 129, no 1, (2003): 53–63.

Jensen, M.E. Empirical methods of estimating or predicting evapotranspiration using radiation. In: Evapotranspiration and Its Role in Water Resources Management. American Society of Agricultural Engineers, Chicago, USA, (1966): 49–53.

Jensen, M.E., and Haise, H.R. Estimating evapotranspiration from solar radiation. J. Irrig. Drain., 89, no 4, (1963): 15–41.

Lu, J., Sun, G., McNulty, S.G., and Amatya, D.M. A comparison of six potential evapotranspiration methods for regional use in the south-eastern United States. J. Am. Water Resour. Assoc., 41, no 3, (2005): 621–633.

Mahringer, W. Verdunstungsstudien am Neusiedler see. Theor. Appl. Climatol., 18, no 1, (1970): 1–20.

Makkink, G.F. Testing the Penman formula by means of lysimeters. J. Inst. Water Eng. Sci., 11, (1957): 277–288.

Mallikarjuna, P., Aruna Jyothy, S., Srinivasa Murthy, D., and Chandrasekhar Reddy, K. Performance of recalibrated equations for the estimation of daily reference evapotranspiration. Water Resour. Manage., 28, (2014): 4513–4535.

Nikam, B.R., Kumar, P., Garg, V., Thakur, P.K., and Aggarwal, S.P. Comparative evaluation of different potential evapotranspiration estimation approaches. Int. J. Res. Eng. Technol, 3, (2014): 543–552.

Priestley, C.H.B., and Taylor, R.J. On the assessment of surface heat flux and evaporation using large scale parameters. Mon. Weather Rev., 100, (1972): 81–92.

Sabziparvar, A.A., Mousavi, R., Marofi, S., Ebrahimipak, N.A., and Heidari, M. An improved estimation of the angstrom—Prescott radiation coefficients for the FAO56 Penman—Monteith evapotranspiration method. Water Resour. Manag., 27, (2013): 2839–2854.

Samaras, D.A., Rei, A., and Theodoropoulos, K. Evaluation of radiation-based reference evapotranspiration models under different Mediterranean climates in central Greece. Water Resour. Manag., 28, (2014): 207–225.

Stephens, J.C. Discussion of "Estimating evaporation from insolation". J. Hydr. Div., 91, no 5, (1965): 171–182.

Stephens, J.C., and Stewart, E.H. A comparison of procedures for computing evaporation and evapotranspiration. In: General Assembly of Berkeley, International Association of Hydrological Sciences, Vol. 62. Berkeley, USA, (1963): 123–133.

Tabari, H., Talaee, P.H., Nadoushani, S.M., Willems, P., and Marchetto, A. A survey of temperature and precipitation based aridity indices in Iran. Quater. Int., 345, (2014): 158–166.

Thejll, P., and Schmith, T. Limitations on regression analysis due to serially correlated residuals: Application to climate reconstruction from proxies. J. Geophys. Res., 110, no D18, (2005): 103.

Trajkovic, S., and Kolakovic, S. Evaluation of reference evapotranspiration models under humid conditions. Water Resour. Manag., 23, no 14, (2009): 3057–3067.

Turc, L. Estimation of irrigation water requirements, potential evapotranspiration: A simple climatic formula evolved up to date. Ann. Agron., 12, (1961): 13–14.

Wen, L. Reconstruction natural flow in a regulated system, the Murrumbidgee River, Australia, using time series analysis. J. Hydrol., 364, (2009): 216–226.

WMO measurement and estimation of evaporation and evapotranspiration. In: Technical Paper (CIMO-Rep.) No. 83. Genova, Switzerlands. (1966).

Xu, C.-Y., Gong, L., Jiang, T., Chen, D., and Singh, V.P. Analysis of spatial distribution and temporal trend of reference evapotranspiration and pan evaporation in Changjiang (Yangtze River) catchment. J. Hydrol., 327, (2006): 81–93.

Xu, C.Y., and Singh, V.P. Evaluation and generalization of temperature-based methods for calculating evaporation. Hydrol. Process., 15, no 2, (2000): 305–319.

Xu, C.-Y., and Singh, V.P. Cross comparison of empirical equations for calculating potential evapotranspiration with data from Switzerland. Water Resour. Manag., 16, (2002): 197–219.

Xu, C.Y., Singh, V.P., Chen, Y.D., and Chen, D. Evaporation and evapotranspiration. In: Singh, V.P. (ed.). Hydrology and Hydraulics, 1st ed. Water Resources Publications, USA, (2008): 229–276.

Yin, Y., Wu, S., Zheng, D., and Yang, Q. Radiation calibration of FAO56 Penman—Monteith model to estimate reference crop evapotranspiration in China. Agric. Water. Manag., 95, (2008): 77–84.

10 Partitioning Models

The partitioning models are divided into mechanistic, empirical, analytical and numeric models. Shuttleworth and Wallace, (1985) defined an analytical model (SW), one for crop (PMc) and for soil surface (PMs):

$$\lambda ET = C_c PM_c + C_s PM_s \tag{206}$$

$$PM_c = \frac{s(R_n - G) + \left[\rho c_p (VPD) - sr_a^c \left(R_n^s - G \right) \right] / \left(r_a^a + r_a^c \right)}{s + \gamma \left[1 + \dfrac{r_s^c}{r_a^a + r_a^c} \right]} \tag{207}$$

$$PM_s = \frac{s(R_n - G) + \left[\rho c_p (VPD) - sr_a^s \left(R_n - G \right) \left(R_n^s - G \right) \right] / \left(r_a^a + r_a^s \right)}{s + \gamma \left[1 + \dfrac{r_s^s}{r_a^a + r_a^s} \right]} \tag{208}$$

λET = [W/m^2]; Cc = canopy surface coefficient; Cs = soil surface coefficient; Cc and Cs are both functions of s, γ, r_a^a, r_a^c, r_s^c, r_a^s and r_s^s; s = rate of change of saturated vapor pressure with air temperature [kPa/°C]; c_p = specific heat of moist air [J/kg °C]; VPD = vapor pressure deficit [kPa]; all measured at a reference height. The terms added to FAO56PM were Rn^s = net radiation at the soil surface [W/m^2]; r_a^a = aerodynamic r between canopy and reference height; r_a^c = bulk boundary r; r_s^c = bulk stomatal r; r_a^s = aerodynamic r between soil surface and canopy; r_s^s = soil surface r, all are in s/m. This model is difficult to parameterize, but it is accurate, so that in many cases, it is simpler.

Lascano et al., (1987) created a model based on energy and water balance; the equations for E and T are based on soil characteristic, plant and atmospheric inputs. The latent heat from soil:

$$\lambda E = \frac{\lambda (AH_s - AH_a)}{r_a} \tag{209}$$

λE = [W/m^2]; AH = absolute humidity [kg/m^3]; subscript s is for soil and subscript a is for air; r_a = aerodynamic resistance [s/m].

Evett and Lascano (1993) defined the latent heat of transpiration as

$$\lambda T = \frac{\lambda (AH_l - AH_a)}{r_{plant}} \tag{210}$$

DOI: 10.1201/9781003467229-10

$\lambda T = W/m^2$; AH_1 = leaf absolute humidity [kg/m³]; r_{plant} = plant resistance formula [s/m].

$$\lambda T = \left(\Psi_s + \Psi_{c\_max} - \Psi_c \right) 1000 \lambda \left(\frac{LAI}{r_{plant\_hyd}} \right) \qquad (211)$$

ψ_s = soil water potential; ψc = water potential; ψ_{c_max} = maximum water potential of the canopy; $r_{plant-hyd}$ = hydraulic resistance of the plant.

Many studies have been carried out using this model with good agreement between the simulated and measured values; ET shows lower values for daily measurements perhaps due to the dry topsoil layer with errors in surface temperature measurement and energy balance calculation. Kool et al., (2014) reported the SWEAT model, proposed by Daamen and Simmonds, (1994), which considers the soil, water, energy and transpiration, not the soil resistance parameter. The approach is based on two layers for E and T interaction; the two layers represent soil and leaves, and canopy is not considered.

$$\lambda ET = \frac{\lambda \left(AH_c - AH_a \right)}{r_a} \qquad (212)$$

$$\lambda E = \frac{\lambda \left(AH_s - AH_c \right)}{r_{scan}} \qquad (213)$$

$$\lambda T = \frac{\lambda \left(AH_l - AH_c \right)}{r_{st} + r_{lbl}} \qquad (214)$$

λET = [W/m²]; λT = [W/m²]; λE = [W/m²]; AH_c = absolute humidity of canopy [kg/m³]; AH_a = absolute humidity of air [kg/m³]; AH_s = absolute humidity of soil [kg/m³]; AH_l = absolute humidity of leaf [kg/m³]; r_a = resistance for canopy to atmosphere; r_{scan} = resistance for soil surface to canopy; r_{st} = resistance stomata; r_{lbl} = resistance for leaf boundary layer. This model gave reliable results when LAI > 2.

Norman et al., (1995), Anderson et al., (1997) and Kustas and Norman, (1999) proposed a two-source energy balance model that estimated measured surface temperature using remote sensing and used the surface temperature to estimate λET. This model needs the directional radiometric temperature of the surface, meteorological data and canopy. There are separate equations for soil and canopy energy balances and, solved partitioning R_n between soil and canopy according to LAI.

Temperature inputs:

$$\lambda E = \lambda E + \lambda T \qquad (215)$$

$$\lambda T = \alpha_{PT} f_g \frac{s}{s + \gamma} R_n^c \qquad (216)$$

α_{PT} = PT constant (1.3); f_g = fraction of green vegetation in the canopy, and R_n^c = net radiation at the canopy [W/m²].

This formula assumed transpiration at the potential rate, and λE is calculated as a residual energy balance. If λE is negative, when λT is bigger than λET, the model is adjusted to make $\lambda E = 0$. Kool et al., (2014) suggested that the model is convenient because it requires few parameters and that it could be extensively used with remote sensing data.

Allen et al., (1998) proposed a dual-K_C model to estimate ET in a well-watered crop, considering a multiplication K_C factor with ET_0 (reference). The K_c factor is divided between the plant component K_{cb} and soil component K_e:

$$ET = (K_{cb} + K_e)ET_0 \qquad (129)$$

ET = [mm/d]; ET_0 = [mm/d]; K_{cb} = dimensionless; K_e = dimensionless.

$$K_{cb} = K_{cb(table)} + [0.004(u_2 - 2) - 0.004(RH_{min} - 45)]\left(\frac{h_p}{3}\right)^{0.3} \qquad (217)$$

$$K_e = K_r(K_{c\_max} - K_{cb}) \le f_{ew}K_{c\_max} \qquad (218)$$

$K_{cb(table)}$ = the value found in the literature that has been measured experimentally; u_2 = mean daily wind speed at 2 m height over grass [m/s]; RH_{min} = mean daily minimum relative humidity [%]; h_p = mean plant height [m]; f_{ew} = fraction of soil from which most evaporation occurs; K_r = reduction factor based on soil water availability; K_{c_max} = maximum evaporation based on available energy for ET at the soil surface.

Ferreira et al., (2012) and Kool et al., (2014) reported that this is the most used model, it requires few parameters and that the results are generally good; the disadvantage is that it is an empirical model for a well-defined crop, so it is not always possible to use under different conditions. Kool et al., (2014) reported on HYDRUS-1D (Šimůnek et al., 2008) as a Windows-based model that simulates the movements of water, heat and solute. This model is based on the Richards equation for saturated media and on the convection–dispersion equation for heat and solute; this last equation is based on Fick's law.

Evaporation is considered a water flux, limited by the potential evaporation (E_{pot}), and transpiration is also limited by potential transpiration (Figure 10.1) (Neuman et al., 1975; Kool et al., 2014), so that E is

$$E = -K\frac{\partial h}{\partial x} - K \le E_{pot} \qquad x = L \qquad (219)$$

The surface boundary pressure head is $h_A \le h$ at $x = L$ \qquad (220)

K = unsaturated soil hydraulic conductivity [m/s]; x = spatial coordinate (positive upwards); L = x-coordinate of the soil surface above a certain reference plane (depth of the soil profile) [m]; h_A = minimum pressure head for prevailing soil conditions [m].

E_{pot} and h_A are predefined by the user or calculated as a humidity function. E_{pot} can be estimated as an ET fraction on Beer's law, and the partitioning of potential ET depends on LAI. Transpiration is a function of root water uptake:

$$T = \int_{L_R} S(h, h_\varphi, x)dx = T_{pot} \int_{L_R} \alpha(h, h_\varphi, x)b(x)\,dx \tag{221}$$

L_R = rooting depth [m]; $S(h, h\phi, x)$ = the sink term defined as the volume of water removed from a unit volume of soil per unit time due to plant water uptake; h_φ= osmotic head [m]; $\alpha(h, h\phi, x)$ = a water stress response function ($0 \leq \alpha \leq 1$) where $\alpha = 1$ implies no water stress; $b(x)$ is the normalized water uptake distribution function [1/m], describing spatial variation of the potential extraction term over the root zone.

Kool et al., (2014) stated this method needs to be better validated; it has been only tested on grass grown at laboratory conditions. The authors highlighted that these portioning studies have been carried out with different combinations of measurements, modeling methods, environmental conditions and agricultural/natural settings, but none of them could be considered particularly accurate. Partitioning ET will be fundamental because of water scarcity combined with increasing world populations.

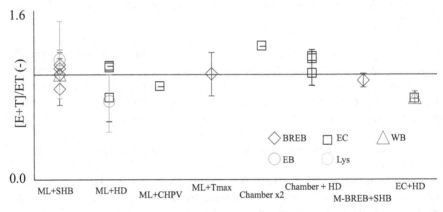

Measurement combination for E and T

FIGURE 10.1 Eight combinations for measuring evaporation from the soil surface (E) and transpiration (T).

Note: The combinations of E and T were compared with evapotranspiration (ET) measurements using the ratio [E + T]/ET with the objective of deriving trends that could indicate the accuracy of the methods. Each point represents a single study. The different colors represent the different methods for estimating ET. Abbreviations: ML: micro-lysimeter, SHB: stem heat balance, HD: heater dissipation, CHPV: compensated heat-pulse velocity, (M)-BREB: (micro) Bowen ratio energy balance, EC: eddy covariance, EB: energy balance, Lys.: weighing lysimeter, WB: water balance.

Source: Modified from Kool et al. (2014).

REFERENCES

Allen, R.G., Pereira, L.S., Raes, D., and Smith, M. Crop evapotranspiration: Guidelines for computing crop water requirements. In: United Nations FAO, Irrigation and Drainage Paper 56. FAO, Rome, Italy. (1998).

Anderson, M.C., Norman, J.M., Diak, G.R., Kustas, W.P., and Mecikalski, J.R. A two-source time-integrated model for estimating surface fluxes using thermal infrared remote sensing. Remote Sens. Environ., 60, (1997): 195–216.

Daamen, C.C., and Simmonds, L.P. Soil Water, Energy and Transpiration—A Numerical Model of Water and Energy Fluxes in Soil Profiles and Sparse Canopies. Department of Soil Science, University of Reading, Reading, UK, (1994).

Evett, S.R., and Lascano, R.J. ENWATBAL.BAS: A mechanistic evapotranspiration model written in compiled basic. Agron. J., 85, (1993): 763–772.

Ferreira, M.I., Silvestre, J., Conceiçao, N., and Malheiro, A.C. Crop and stress coefficients in rain fed and deficit irrigation vineyards using sap flow techniques. Irrig. Sci., 30, (2012): 433–447.

Kool, D., Agama, N., Lazarovitcha, N., Heitmanc, J.L., Sauerd, T.J., and Ben-Gal, A. A review of approaches for evapotranspiration partitioning. Agric. For. Meteorol., 184, (2014): 56–70.

Kustas, W.P., and Norman, J.M. Evaluation of soil and vegetation heat flux predictions using a simple two-source model with radiometric temperatures for partial canopy cover. Agric. For. Meteorol., 94, (1999): 13–29.

Lascano, R.J., Van Bavel, C.H.M., Hatfield, J.L., and Upchurch, D.R. Energy and water balance of a sparse crop: Simulated and measured soil and crop evaporation. Soil Sci. Soc. Am. J., 51, (1987): 1113–1121.

Neuman, S.P., Feddes, R.A., and Bresler, E. Finite element analysis of two-dimensional flow in soils considering water uptake by roots. I. Theory. Soil Sci. Soc. Am. J., 39, (1975): 224–230.

Norman, J.M., Kustas, W.P., and Humes, K.S. A two-source approach for estimating soil and vegetation energy fluxes in observations of directional radiometric surface temperature. Agric. For. Meteorol., 77, (1995): 263–293.

Shuttleworth, W.J., and Wallace, J.S.Evaporation from sparse canopy: An energy combination theory. Q. J. Meterol. Soc., 111, (1985): 839–855.

Šimůnek, J., Šejna, M., Saito, H., Sakai, M., and Van Genuchten, M.T. The HYDRUS-1D software package for simulating the movement of water, heat, and multiple solutes in variably saturated media, version 4.08. HYDRUS Software Series 3. Department of Environmental Sciences, University of California Riverside, Riverside, CA, USA. (2008).

11 ET Machine Learning Models

11.1 ARTIFICIAL NEURAL NETWORKS

Kisi, (2007) calculated ET_0 with artificial neural networks (ANNs) using the Levenberg–Marquardt (LM) algorithm, and the resulting ANNs gave excellent data. Moreover, the HG results were better than those for Penman and Tc.

Abedi-Koupai et al., (2009) compared ANN results using lysimeter and the Stanghellini, PE, Penman–Monteith (PM) and Fynn equations. The study was carried out in the greenhouse at Isfahan University of Technology, and the authors compared the ET_0 lysimeter data according to the following equation:

$$ET_0 = (SWC_{to} - SWC_{tl}) + I - D \qquad (222)$$

$(SWC_{to} - SWC_{tl})$ = change in volumetric soil water content between two measurements; I = total volume of applied irrigation water; D = total volume of collected drainage.

Stanghellini:

$$ET_o = \frac{2LAI}{\lambda} \frac{\Delta(R_n - G) + K_t \dfrac{VPD\rho C_p}{r_a}}{\Delta + \left(1 + \dfrac{r_c}{r_a}\right)} \qquad (223)$$

LAI = leaf area index [m^2/m^2]; K_t = time unit conversion factor, 86,400 [s/day]; Δ = slope of saturation vapor pressure-temperature curve [Pa/°C]; VPD = vapor pressure deficit [kPa]; G = soil heat flux [MJ/m^2day]; R_n = net radiation [MJ/m^2day]; ρ = air density [kg/m^3]; r_c = canopy resistance [s/m]; r_a = aerodynamic resistance [s/m]; λ = latent heat of vaporization of water [MJ/kg]; C_p = air specific heat at constant pressure [MJ/kg°C]. In this equation, the solar radiation heat is calculated from the empirical characteristics of short- and long-wave radiation absorption in a multilayer canopy.

PM (Monteith, 1965):

$$ET_o = \frac{1}{\lambda} \frac{\Delta(R_n - G) + K_t \dfrac{VPD\rho C_p}{r_a}}{\Delta + \left(1 + \dfrac{r_c}{r_a}\right)} \qquad (224)$$

DOI: 10.1201/9781003467229-11

K_t = time unit conversion factor, 86,400 [s/day]. This equation is used to estimate daily or hourly ET_0 for two references surface.

Fynn (1993):

$$ET_o = \frac{2LAI\,\rho C_p \dfrac{(e_s - e_a)}{r_a} + \Delta\left(R_n - G\right)}{\lambda\gamma\,ri_c} \qquad (225)$$

This equation is valid for measuring ET_0 in a greenhouse; it is based on total net radiation. LAI modifies the vapor pressure because all of the layers of the canopy participate in the water vapor exchange.

Abedi-Koupai et al., (2009) proposed an ANN model with four variables: air temperature, relative humidity, solar radiation and wind speed (Figure 11.1). The model was based on two hidden layers and five hidden neurons and used a log sigmoid function. This model provided the best estimation of efficiency, followed by Stanghellini, PM, Fynn and PE. Unfortunately, Fynn produced poor results in the greenhouse. The best ET values were in order: ANNs, PM and PE.

$$EF = \frac{\sum_{i=1}^{n}(O_i - \bar{O})^2 - \sum_{i-1}^{n}(O_i - E_i)^2}{\sum_{i=1}^{n}(O_i - \bar{O})^2} \qquad (226)$$

Later, Eslamian et al., (2012) compared FAO56PM with ANNs and an ANN genetic algorithm (ANN-GA). The networks were based on an input layer, hidden layer(s) and the output layer, where the results are produced. This ANN is called a multilayer perceptron (Fausset, 1994). Eslamian et al., (2012) also

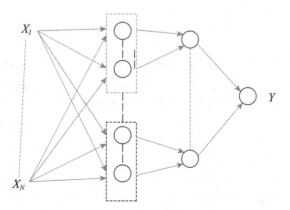

FIGURE 11.1   Schematic of the proposed ANN model.

Source: Modified from Eslamian et al. (2012).

tested ANN-GA, which first initialized the connection weights and the input layer nodes and then calculated the global error at the output layer as the suitable function. These processes were repeated to reach the optimal solution (Figure 11.2, Table 11.1).

Eslamian et al., (2012) used five input variables to test the algorithm: maximum air temperature, minimum air temperature, average air temperature, relative humidity, sunshine and wind speed, and they used FAO56PM to calculate monthly ET_0. In ANNs, the hidden nodes are determined by trial and error; inputs and outputs are normalized to improve the performance (Table 11.2).

$$xn_{i,k} = \frac{x_{i,k} - m_k}{SD_k} \tag{227}$$

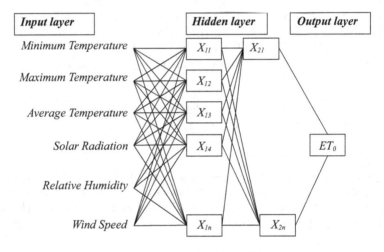

FIGURE 11.2 Schematic of the proposed ANN-GA model.

Source: Modified from Eslamian et al. (2012).

TABLE 11.1
Specifications and Optimal Structure of the Proposed ANN-GA Model

Learning rule	Number of neutrons	Threshold function	Epoch	Momentum coefficient	Learning coefficient	Final training (repeat)	Validation error
Quick prop	8.2	Sigmoid	12	0.4	0.25	10000	**0.0438**
Delta rule	6.4	Tan H	6	0.6	0.5	5000	**0.0811**
Norm. Cum. D	6.4	Sigmoid	16	0.4	0.75	14000	**0.0826**
Max prop	8.2	Tan H	10	0.8	1	16000	**0.0851**
Delta. Bar. De	9.1	Tan H	24	0.4	0.75	20000	**0.0907**
Ext DBD	10.1	Sigmoid	18	0.6	0.5	10000	**0.0966**

Source: Modified from Eslamian et al. (2012).

TABLE 11.2
The Model Errors for Different Percentages of Input Data

Model	Test error	Validation error	Training error	Percent of training data	Percent of validation data (%)
ANNs	0.0488	0.0397	0.0960	60	20
	0.1071	0.0818	0.6720	50	25
ANN-GA	0.0677	0.0434	0.109	60	20
	0.188	0.797	0.0655	50	25

Source: Modified from Eslamian et al. (2012).

$xn_{i,k}$ = normalized input K or target data at i = 1,2 . . . N; $x_{i,k}$ = original data; m_{kv} = mean value of input k; SD_{kv} = standard deviation of input k.

The authors estimated the ANNs and the Gas ET model with the coefficient of determination (R):

$$R^2 = \frac{E_o - E}{E_o} \tag{228}$$

The authors determined that humidity, minimum air temperature and sunshine were the sensitive variables. Garson, (1991) found that ANNs well estimated ET_0 with limited climate data, and this was confirmed by Sudheer et al., (2003); Trajkovic et al., (2003); Odhiambo et al., (2001) and Eslamian et al., (2012).

Eslamian et al., (2012) identified good relationships between ET_0 from FAO56PM and from the ANNs and ANN-GA; both of the latter well estimated monthly ET_0 using climatic data. In particular, the ANNs and ANN-GA overcame the linear problems, and the authors recommended using this method in combination with the neuro-fuzzy model.

Goodarzi and Eslamian, (2018) tested linear and nonlinear models for estimating reference ET. They calculated radial basis function (RBF) neural networks and genetic programming (GP) as the nonlinear models and multiple linear regression (MLR) for the linear model. Goodarzi and Eslamian, (2018) highlighted the differences among the ANN types based on the node numbers and, how functions are calculated. In particular, RBF is a nonlinear hybrid network with a single hidden layer; the transfer function is the middle layer (Gaussian function), and the output layer is a linear function. The hidden layer neurons attach to input neurons through weight parameters and the output of this layer is a function of distance between the input and radial vectors:

$$\delta = \sqrt{\sum_{i=1}^{n}\left(x_i - x_{ij}\right)} \tag{229}$$

The output layer is a Gaussian function described by

$$f(\delta_i) = Exp(-\lambda\delta_i^2) \tag{230}$$

λ = constant coefficient.

The equation of the outputs of output layer is

$$z_k = \sum_{j=1}^{j} b_{jk} y_j \tag{231}$$

b_{jk} = weight coefficients between j^{th} hidden layer neurons and k^{th} output layer neuron; $y_j = j^{th}$ output of the hidden layer neurons.

The RBF method of selecting a suitable network is based on trial and error to obtain the optimal RMSE and (R^2):

$$RMSE = \sqrt{\frac{1}{N}\sum_{i=1}^{N}(y_t - y_o)^2} \tag{232}$$

$$R^2 = 1 - \frac{\sum y_t - y_o}{\sum y_t^2 - \dfrac{\sum y_o^2}{n}} \tag{233}$$

Goodarzi and Eslamian, (2018) configured eight different RBF models for identifying ET_0 configurations by changing the architecture and parameters. Table 11.3 shows the testing results.

TABLE 11.3
Results from Testing Different RBF Models

Model	Input parameters	Optimized architecture	Testing R²	Testing RMSE	Training R²	Training RMSE
RBF1	T_{max}, T_{min}, W, H, R_n	(5–12–1)	0.96	0.49	0.96	0.46
RBF2	T_{max}, T_{min}, W, R_n	(4–9–1)	0.95	0.61	0.95	0.51
RBF3	T_{ave}, W, H, R_n	(4–9–1)	0.94	0.54	0.94	0.58
RBF4	T_{max}, T_{min}, W, H	(4–9–1)	0.93	0.61	0.93	0.65
RBF5	T_{max}, T_{min}, W	(3–9–1)	0.92	0.75	0.92	0.78
RBF6	T_{max}, T_{min}, R_n	(3–9–1)	0.91	0.88	0.92	0.80
RBF7	T_{max}, T_{min}, H	(3–9–1)	0.90	0.85	0.91	0.87
RBF8	T_{max}, T_{min}	(3–9–1)	0.91	0.92	0.90	0.97

Source: Modified from Goodarzi and Eslamian (2018).

The authors calculated the sensitivity for each parameter (maximum temperature, minimum temperature, average temperature, RH, wind speed and radiation) as well as K_{Sp}. From these eight models, Goodarzi and Eslamian, (2018) determined that average temperature was the most sensitive parameter and RH the least for each method. The finding that temperature had the most significant influence on ET confirmed results from Huo et al., (2012) and Jain et al., (2008). Figure 11.3 displays a scatter plot of the findings for the eight RBF models compared with PM. Figure 11.4 shows the comparison findings for one of the models, RBF1.

Genetic programming (GP) is an evolutionary algorithm based on Darwin's theonatural selection and genetics (Koza, 1992; Khu et al., 2001). This algorithm defines an objective function as a quality criterion, comparing the solutions to apply the correct data structure. First, the fitting function is selected according to RMSE. Second, all of the total input variables and function must be selected; in this case, the input variables are meteorological data (Babovic and Keijzer, 2000; Parasuraman et al., 2007). Table 11.4 shows the results for all eight GP models against PM.

Figure 11.5 displays the scatter plot comparing the GP model results with PM. Figure 11.6 displays the scatter plot for comparing the predicted ET from model GP3 with the PM results.

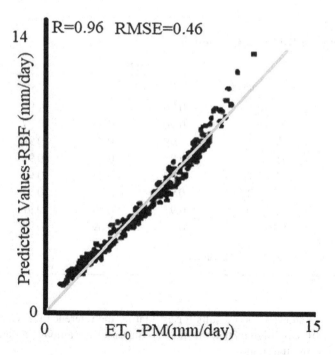

FIGURE 11.3 Scatter plot of the RBF method compared with PM.

Source: Modified from Goodarzi and Eslamian (2018).

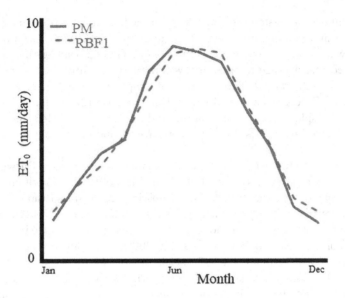

FIGURE 11.4 ET_0 predicted by RBF1 compared with PM for 2002.

Source: Modified from Goodarzi and Eslamian (2018).

TABLE 11.4
Comparing the Different GP Models

Model	Input parameters	Testing		Training	
		R²	RMSE	R²	RMSE
GP1	T_{ave}, W, R_n	0.96	0.42	0.96	0.38
GP2	T_{ave}, W, H	0.96	0.43	0.97	0.35
GP3	T_{ave}, W, H, N	0.99	0.21	0.99	0.24
GP4	T_{ave}, H, R_n	0.92	0.65	0.92	0.70
GP5	T_{max}, T_{min}, W	0.95	0.49	0.96	0.42
GP6	T_{max}, T_{min}, R_n	0.91	0.84	0.92	0.8
GP7	T_{max}, T_{min}, H	0.92	0.75	0.91	0.74
GP8	W, H, R_n	0.92	0.69	0.93	0.65

Source: Modified from Goodarzi and Eslamian (2018).

Although GP3 showed good results for estimating the reference ET, Goodarzi and Eslamian, (2018) highlighted that the model requires a large amount of data and does not work well in other regions.

MLR models are based on two independent variables and one dependent variable fitting a linear equation to the observed data. The relationships among all these variables are not well known, and this allows for the creation of a predictive model.

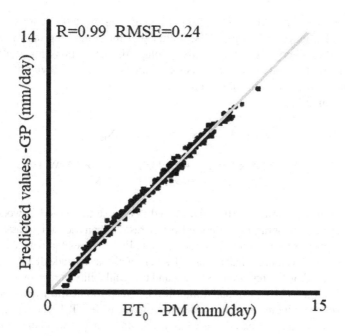

FIGURE 11.5   Scatter plot of the GP model findings compared with PM.

Source: Modified from Goodarzi and Eslamian (2018).

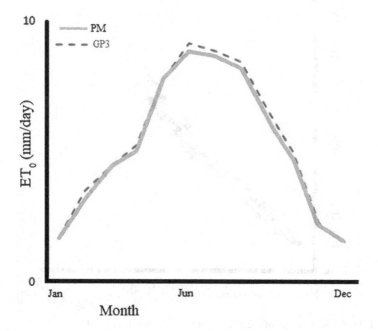

FIGURE 11.6 ET_0 predicted by GP3 compared with PM for 2002.

Source: Modified from Goodarzi and Eslamian (2018).

Meteorological data are the independent variables, and the dependent variable is the reference ET. Figure 11.7 displays the scatter plot for comparing the MLR predictions with PM. Figure 11.8 graphically displays the predictions for model MLR1 compared with PM. Table 11.5 presents the findings for the sensitivity of each of the weather parameters in terms of each of the three study models.

MLR format:

$$Y = \beta_0 + \beta_1 X_{i1} + \beta_2 X_{i2} + \ldots \beta_p X_{ip} \qquad (234)$$

$$ET_0 = 5.05 \times 10^{-2} T_{min} + 0.1235 T_{min} + 6.6 \times 10^{-3} W + 0.12861 R_n - 3.39 \times 10^{-3} H - 2.168 \qquad (235)$$

Goodarzi and Eslamian, (2018) highlighted that these three models work well; in particular, genetic programming showed better results than the radial basic function, but RBF is simpler to use. Moreover, the multiple linear regression data showed good accuracy, but RFB is more suitable than MLR. The authors conclude that models should be selected based on the researcher's needs and the availability of meteorological data.

Landeras et al., (2008) evaluated multilayer perceptron neural networks combining meteorological inputs and one hidden layer to measure ET_0 based on daily values from FAO56PM; they compared these data with data from four weather stations. They

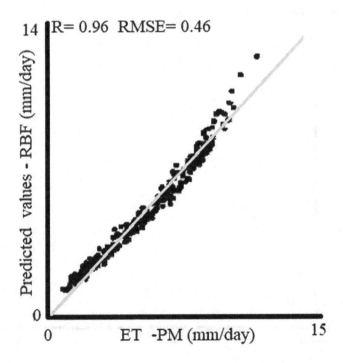

FIGURE 11.7 Scatter plot of the MLR method compared with PM.

Source: Modified from Goodarzi and Eslamian (2018).

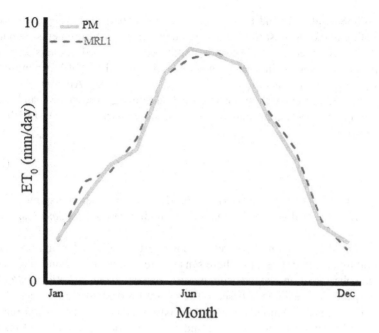

FIGURE 11.8 ET_0 predicted by MLR1 compared with PM for 2002.

Source: Modified from Goodarzi and Eslamian (2018).

TABLE 11.5
Sensitivity of the Reference RT Parameters under the Three Prediction Models

Parameter	RBF	GP	MLR
T_{max}	0.54	0.56	0.65
T_{min}	0.37	0.37	0.41
T_{ave}	0.72	0.88	0.82
W	0.29	0.34	0.58
H	−0.03	−0.08	−0.015
R_n	0.64	0.48	0.68

Source: Modified from Goodarzi and Eslamian (2018).

used Intelligent Problem Solver (IPS) and the integrated algorithms to select the ANN architecture. The advantage in using the IPS module is that it experimentally tests the network configurations using many algorithms. In particular, they highlighted the ANNs would support a model with the same meteorological features as ET_0.

Landeras et al., (2008) used IPS to iteratively run the outputs from each network through training patterns to create a multidimensional error function using four algorithms: HS, Makk, Tc and PT. These variants are characterized by their use of solar radiation and/or RH estimation and not temperature measurements. Landeras et al., (2008) compared the ANNs with 10 ET_0 equations and variants in Spain. Researchers recommend using these equations rather than FAO56PM when meteorological parameters are lacking (Droogers and Allen, 2002). Allen et al., (1994) indicated the need to calibrate empirical methods through FAO56PM, and in the absence of, they suggest calibrating alternative equations with FAO56PM with data from a near station:

$$ET_0 = a + bE_{model} \tag{236}$$

ET_0 = reference evapotranspiration (FAO56PM); E_{Model} = evapotranspiration estimated by any of the evaluated ET_0 equations; a and b = regression constants.

Landeras et al., (2008) created three groups to estimate ET. The first was based on temperature and/or RH because these sensors are in weather stations; the second group used solar radiation methods and the third one used solar radiation and/or RH and wind speed. The equations based on temperature data showed high RMSEs, and HS had bad results. Among the equations based on temperature and/or relative humidity, Makk, based on temperature and solar radiation, showed good results. Tc, based on solar radiation, showed the best performance because it introduced a correction for RH, but Tc showed low values at all of the stations except for RMSE at Arkaute station. At this station, Makk showed low RMSEs, and the solar radiation equations showed high errors; the solar radiation equations gave low errors at Zambrana station, where daily RH is always very low:

$$MBE = \frac{\sum_{i=1}^{n}\left(E_{Modeli} - E_{PM\,56i}\right)}{n} \tag{237}$$

$$MAE = \left| \frac{\sum_{i=1}^{n}\left(E_{modeli} - E_{PM\,56i}\right)}{n} \right| \tag{238}$$

$$RMSE = \sqrt{\frac{\sum_{i=1}^{n}(E_{modeli} - E_{PM\,56i})^2}{n}} \tag{239}$$

FAO56PM had the lowest RMSEs, particularly at Salvatierra and Navarrete stations. Landeras et al., (2008) observed that errors from solar radiation methods were lower than the values from RH methods. The different error values from Arkaute and Zambrana stations seem to depend on the positive correlations between the errors in Makk and Tc and RH and the negative correlation with the wind speed. See Table 11.6 for the comparison results.

Table 11.7 shows that HS estimation errors correlated positively with ΔT and solar radiation, explaining the high error values at Zambrana station. Conversely, groups

TABLE 11.6
Comparison of ANNs and Calibrated ET_0 Equations under Different Meteorological Data Requirements

Model	Inputs	Meteorological data requirements
	Temperature and/or RH-based methods	
NN1	$T_{min, max, mean}$, R_a	$T_{min, max, mean}$
$E_{PTRsest}$	$T_{min, max, mean}$	$T_{min, max, mean}$
$E_{MKRsetst}$	$T_{min, max, mean}$	$T_{min, max, mean}$
E_{HS}	$T_{min, max, mean}$, R_a	$T_{min, max, mean}$
NN3	$T_{min, max, mean}$, HR, R_a	$T_{min, max, mean}$, HR
E_{TRsest}	$T_{min, max, mean}$, HR, R_a	$T_{min, max, mean}$, HR
	Solar radiation methods	
NN2	$T_{min, max, mean}$, R_s, R_a	$T_{min, max, mean}$, R_s
E_{PT}	$T_{min, max, mean}$, R_s	$T_{min, max, mean}$, R_s
E_{MK}	$T_{min, max, mean}$, R_s	$T_{min, max, mean}$, R_s
NN4	$T_{min, max, mean}$, HR, R_s, R_a	$T_{min, max, mean}$, HR, R_s
E_T	$T_{min, max, mean}$, HR, R_s	$T_{min, max, mean}$, HR, R_s
	FAO56PM analogous methods	
NN5	$T_{min, max, mean}$, u_2, HR, R_a	$T_{min, max, mean}$, u_2, HR
$E_{PM56Rsest}$	$T_{min, max, mean}$, u_2, HR, R_a	$T_{min, max, mean}$, u_2, HR
NN6	$T_{min, max, mean}$, u_2, R_s, R_a	$T_{min, max, mean}$, u_2, R_s
$E_{PM56Hsest}$	$T_{min, max, mean}$, u_2, R_s	$T_{min, max, mean}$, u_2, R_s
NN7	$T_{min, max, mean}$, u_2, R_a	$T_{min, max, mean}$, u_2
$E_{PM56RsHRest}$	$T_{min, max, mean}$, u_2, R_a	$T_{min, max, mean}$, u_2

Source: Modified from Landeras et al. (2008).

TABLE 11.7
Parameter Inputs for Each ANN

	NN1	NN2	NN3	NN4	NN5	NN6	NN7
T_{mean}	y	y	y	y	y	y	y
T_{max}	y	y	y	y	y	y	y
T_{min}	y	y	y	y	y	y	y
R_a	y	y	y	y	y	y	y
R_s		y		y		y	
HR			y	y	y		
u_2					y	y	y

Note: T_{mean} = mean temperature; T_{max} = maximum temperature; T_{min} = minimum temperature; R_a = extraterrestrial radiation; R_s = solar radiation; HR = relative humidity; u_2 = wind speed at 2 m height.

Source: Modified from Landeras et al. (2008).

based on the temperature showed negative correlations with wind speed, confirming the overestimation at Arkaute station due to the low wind speed. Landeras et al., (2008) compared daily ET_0 calibrated with FAO56PM with the ANN values. The equations with the highest improvements after the calibration were HS and PT with estimated solar radiation. In contrast, Tc and Makk did not show improvements at Salvatierra or Navarrete, Tc did not improve the errors at Zambrana and Makk did not improve findings at Arkaute.

Landeras et al., (2008) observed that NN4 showed excellent results, in contrast with the models based on RH, temperature or solar radiation. NN3 and NN1 showed good results with the models based on temperature and RH. Among the equations calibrated with ET_0 equations, HS produced the lowest RMSE. The authors confirmed that ANNs performed better than the calibrated ET_0 equations and among the calibrated equations recommended using ANNs at locations that lack suitable meteorological data. PT performed better than the other calibrated equation because local calibration produces better results.

In terms of specific models, Landeras et al., (2008) recommend NN1 if only temperature data are available and NN2 when solar radiation and temperature data are available. They recommend NN4 if temperature, solar radiation and RH data are available. Additionally, they call for NN5 with available RH data, NN6 with solar radiation data and NN7 with wind speed and temperature. Table 11.7 presents the parameter inputs necessary for each network.

11.2   FUZZY LINEAR REGRESSION

Tabari et al., (2012) studied the ANFIS neurofuzzy inference system for estimating crop evapotranspiration (ET_C), comparing ANFIS and support vector machine (SVM) with BC, Makk, Tc, PT, HG and the Ritchie equation. SVM uses temperature, RH, solar radiation, sunshine hours, and wind speed. ET_C can be calculated by measuring ET_0 and applying a crop coefficient (K_C), and according to the authors, ANNs are ideal for creating valid ET_C estimation models.

The ANFIS model gave better results than the other methods (Tabari et al., 2012). For their research, Tabari et al., (2012) applied ANFIS in a province in Iran (Eastern Azarbayjan province). They used FAO56PM to compare daily ET_0 and examined four periods to validate this model for a potato crop: initial period, crop's growth and expansion period, middle period and final period. The authors applied three different K_Cs. They modified the K_Cs for the middle and final periods to consider local wind speed, RH and potato maximum height at the site ($K_{C\text{-mid-table}}$ and $K_{C\text{-end-table}}$).

Neurofuzzy systems are based on the semantic transparency of rule-based fuzzy systems with neural networks (Kurtulus and Razack, 2010; Tabari et al., 2012). Combining ANNs and fuzzy logic speed improves fault tolerance and adaptiveness (Setlak, 2008; Tabari et al., 2012). Neurofuzzy models are multilayer neural network-based fuzzy systems with five layers; the input and output nodes are the input states and output response. In the hidden layer, there are nodes such as the membership function and rules, avoiding the problem that normal feed-forward multilayer networks are difficult to understand and modify.

ANFIS is based on a hybrid-learning algorithm with the combination of gradient descent and least square methods (see Figure 11.9). The aim is to define the nodes as the membership function parameters and consequences parameters so that ANFIS response and network response match the training data. Two membership functions are reliable for ET_C. Temperature is highly correlated with ET_C, data showed that the model could be improved by adding aerodynamic effects such as RH, which shows a negative relationship with ET_C. Solar radiation and wind speed showed positive relationships with the results from ET_C. Tabari et al., (2012) tested five input combinations: air temperature; air temperature + relative humidity; air temperature + relative humidity + solar radiation; air temperature + relative humidity + solar radiation + sunshine hours; air temperature + relative humidity + solar radiation + sunshine hours + wind speed. Model ANFIS5 gave good results that agreed with FAO56PM, but ANFIS1 showed the highest error rate. All the other equations show low ET_Cs determined by FAO56PM except for BC.

Fuzzy regression is based on the concept that residuals between estimators and observations are not caused by mistakes but derive from the uncertainty parameter; therefore, the distribution deals with real observation (Amiri et al., 2019). Amiri et al., (2019) applied fuzzy regression on a greenhouse and used a lysimeter to measure the soil–water balance to construct a reference for ET_0:

$$ET_0 = (SWC_{to} - SWC_{tl}) + I - D \qquad (222)$$

$(SWC_{to} - SWC_{tl})$ = change in volumetric soil water content between two measurements dates; I = total volumes of applied irrigation water; D = collected drainage for the period under consideration. Tensiometers control the application of water at the study site; they are installed at 15 cm depth. When soil water potential was at -20KPa at depth of 15 cm, irrigation started.

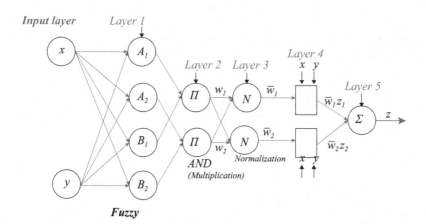

FIGURE 11.9 ANFIS architecture.

Source: Modified from Tabari et al. (2012).

Notes: In the scheme, a circle is a fixed node, and a square is an adaptive node. A_i and B_i are the fuzzy sets in the antecedent.

The equation for fuzzy linear regression (FLR) is:

$$\tilde{Y} = \tilde{A}_0 + \tilde{A}_1 x_{i1} + \ldots + \tilde{A}_n x_{in} = \tilde{A} x_i \tag{240}$$

$x_i = [x_0, x_{i1}, \ldots, x_{in}]$ is a vector of independent variables in the ith data; $I = 1, \ldots, m$;
$\tilde{A} = [\tilde{A}_0, \ldots, \tilde{A}_n]$ is a vector of fuzzy parameters exhibited in the form of symmetric triangular fuzzy numbers denoted by $\tilde{A}_j = (p_j, c_j)$, $j = 1, \ldots, n$ where pj is its central value and cj is its half width.

The regression coefficients are fuzzy numbers, and the dependent variable $\tilde{Y}$ is also a fuzzy number (Figure 11.10).

The linear formula is as follows; Figure 11.11 graphically displays fuzzy linear relationships:

$$\tilde{Y} = (p_0, c_0) + (p_1, c_1) x_{i1} + \ldots + (p_n, c_n) x_{in} \tag{241}$$

For this FLR, the weather data were maximum, minimum and average temperature; maximum, minimum and average RH; wind speed and solar radiation (Amiri et al., 2019). Amiri et al., (2019) used some of these data for the FLR and the

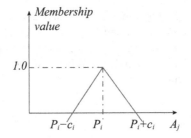

FIGURE 11.10   Triangular representation of fuzzy numbers.

Source: Modified from Amiri et al. (2019).

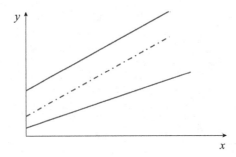

FIGURE 11.11   A fuzzy linear relationship.

Source: Modified from Amiri et al. (2019).

remainder for the lysimeter equation, having observed that some parameters were more important than the others. They selected the most significant parameters for the final models. They obtained three models:

$$ET_0 = (0.5431, 0) + (0.1733, 0.1271) \text{ Tmin} + (0.08652, 0) \text{ Tmax} + (0, 0.03544) \text{ RHmean} + (0.0588, 0) \text{ U} + (0.165, 0) \text{ R} \tag{242}$$

$$ET_0 = (0.760) + (0.1833, 0.1684) \text{ Tmin} + (0.089, 0) \text{ Tmax} + (0, 0.1653) \text{ RHmin} + (0,0) \text{ RHmax} + (0.1975, 0) \text{ u} + (0.1938, 0.0847) \text{ R} \tag{243}$$

$$ET_0 = (0.0734, 0) + (0.1637, 0.1356) \text{ Tmean} + (0.00689, 0.239) \text{ RHmin} + (0, 0) \text{ RHmax} + (0.3668, 0.093) \text{ u} + (0.1813, 0) \text{ R} \tag{244}$$

Using meteorological data, the ET rate is a fuzzy number, so that using the center method it becomes a number. Amiri et al., (2019) found that the most suitable model was the first one, considering the maximum and minimum temperature, mean RH, wind speed and solar radiation. The first model produced a RMSE of 0.68 mm/day and R^2 of 0.98 (Figure 11.12). Amiri et al., (2019) identified wind speed and solar radiation as the most sensitive variables:

$$RMSE = \sqrt{\frac{1}{N} \sum_{i=1}^{n} (X_k - Y_k)^2} \tag{245}$$

$$R^2 = \left[\frac{\sum_{K=1}^{n} (X_k - \bar{X})(Y_k - \bar{Y})}{\sum_{K=1}^{n} (X_k - \bar{X})^2 \sum_{Y=1}^{N} (Y - \bar{Y})^2} \right]^2 \tag{246}$$

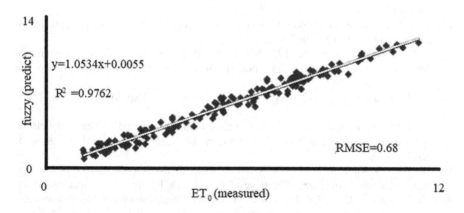

FIGURE 11.12   Comparison between observed and predicted (fuzzy 1) ET_0.

Source: Modified from Amiri et al. (2019).

TABLE 11.8
Comparison of Fuzzy Linear Regression Model Results with ET from Lysimeter Data

Method	Model	Performance	RMSE	R^2
Fuzzy 1	$ET_{lysimeter}-0.949ET_{fuzzy\ 1}$	Very good	0.68	0.98
Fuzzy 2	$ET_{lysimeter}-0.77ET_{fuzzy\ 2}$	Reasonable	0.99	0.94
Fuzzy 3	$ET_{lysimeter}-0.71ET_{fuzzy\ 3}$	Not good	1.30	0.92

Source: Modified from Amiri et al. (2019).

Table 11.8 presents the full comparison findings. The first model provided the best estimated ETs with the lowest RMSEs, and it overestimated ET_0 only by 5%.

REFERENCES

Abedi-Koupai, J., Amiri, M.J., and Eslamian, S.S. Comparison of artificial neural network and physically based models for estimating of reference evapotranspiration in greenhouse. Aust. J. Basic Appl. Sci., 3, no 3, (2009): 2528–2535.

Allen, R.G., Smith, M., and Pereira, L.S. An update for the definition of reference evapotranspiration. ICID Bull., 43, (1994): 1–34.

Amiri, M.J., Zarei, A.R., Abedi-Koupai, J., and Eslamian, S.T. The performance of fuzzy regression method for estimating of reference evapotranspiration under controlled environment. Int. J. Hydrol. Sci. Tech., 9, no 1, (2019): 28–38.

Babovic, V., and Keijzer, M. Genetic programming as model induction engine. J. Hydroinform., 2, no 1, (2000): 35–60.

Droogers, P., and Allen, R.G. Estimating reference evapotranspiration under inaccurate data conditions. Irrig. Drain. Syst., 16, (2002): 33–45.

Eslamian, S.S., Gohari, S.A., Zareian, M.J., and Firoozfar, A. Estimating Penman—Monteith reference evapotranspiration using artificial neural networks and genetic algorithm: A case study. Arab. J. Sci. Eng., 37, (2012): 935–944.

Fausset, L.V. Fundamentals of Neural Networks: Architectures, Algorithms and Applications. Prentice Hall, Upper Saddle River, (1994).

Fynn, R.P., A. Al-Shooshan, T.H. Short and R.W. McMahon. Evapotranspiration Measurements and Modeling for a Potted Chrysanthemum Crop. Transactions of the ASAE, 36, no 6, (1993): 1907–1913.

Garson, G.D. Interpreting neural-network connection weights. Exp. Syst. Appl., 6, (1991): 47–51.

Goodarzi, M., and Eslamian, S. Performance evaluation of linear and nonlinear models for the estimation of reference evapotranspiration. Int. J. Hydrol. Sci. Technol., 8, no 1, (2018): 1–15.

Huo, Z., Feng, S., Kang, S., and Dai, X. Artificial neural network models for reference evapotranspiration in an arid area of northwest China. J. Arid Environ., 82, (2012): 81–90.

Jain, S.K., Nayak, P.C., and Sudheer, K.P. Models for estimating evapotranspiration using artificial neural network, and their physical interpretation. Hydrol. Process., 22, no 13, (2008): 2225–2234.

Khu, S.T., Liong, S.Y., Babovic, V., Madsen, H., and Muttil, N. Genetic programming and its application in real-time runoff forming. J. Am. Water Resour. Assoc., 37, no 2, (2001): 439–451.

Kisi, O. Evapotranspiration modeling from climatic data using a neural computing technique. Hydrol. Process., 21, (2007): 1925–1934.

Koza, J.R. Genetic Programming: On the Programming of Computers by Means of Natural Selection. The MIT Press, Cambridge, MA, USA, (1992).

Kurtulus, B., and Razack, M. Modeling daily discharge responses of a large karstic aquifer using soft computing methods: Artificial neural network and neuro-fuzzy. J. Hydrol., 381, (2010): 101–111.

Landeras, G., Ortiz-Barredo, A., and López, J.J. Comparison of artificial neural network models and empirical and semi-empirical equations for daily reference evapotranspiration estimation in the Basque Country (Northern Spain). Agric. Water Manag., 95, no 5, (2008): 553–565.

Monteith, J.L. Evaporation and environment. In: Fogg, G.E. (ed.). The State and Movement of Water in Living Organisms., Proc. Symp. Soc. Exp. Biol. Academic Press, (1965): 205–234.

Odhiambo, L.O., Yoder, R.E., Yoder, D.C., & Hines, J.W. Optimization of fuzzy evapotranspiration model through neural training with input–output examples. Transactions of the ASAE, 44, no 6, (2001): 1625.

Parasuraman, K., Elshorbagy, A., and Carey, S.K. Modeling the dynamics of the evapotranspiration using genetic programming. Hydrol. Sci. J., 52, no 3, (2007): 563–578.

Setlak, G. The fuzzy-neuro classifier for decision support. Int. J. Inform. Theor. Appl., 15, (2008): 21–26.

Stanghellini, C. Transpiration of greenhouse crops: An aid to climate management. Ph.D. thesis, Wageningen Agricultural Univ., Wageningen, The Netherland. (1987).

Sudheer, K.P., Gosain, A.K., and Ramasastri, K.S. Estimating actual evapotranspiration from limited climatic data using neural computing technique. J. Irrig. Drain. Eng. ASCE, 129, (2003): 214–218.

Tabari, H., Grismer, M.E., and Trajkovic, S. Comparative analysis of 31 reference evapotranspiration methods under humid conditions. Irrig. Sci., 31, (2012): 107–117.

Trajkovic, S., Todorovic, B., and Stankovic, M. Forecasting reference evapotranspiration by artificial neural networks. J. Irrig. Drain. Eng., 129, no 6, (2003): 454–457.

12 ET in a Rain Garden

Hess et al., (2019) applied two ET equations in a rain garden; the data were from lysimeters. The aim was to find a tool for incorporating an ET equation into stormwater control measures. The authors used FAO56MP and Hargreaves equation; ET data were compared and calibrated from both equations, considering or not the reduction factor for water availability (like moisture) and a factor for crop development.

Hess et al., (2019) modeled water availability using the soil moisture extraction function (SMEF), which is a function of soil moisture and field-capacity soil moisture; it reflects the relationship between the water content and maximum water-holding capacity. Crop coefficients (K_c) represent plant growth, starting with K_{cini} during the first stage. K_{cmid} in the middle/development stage is interpolated from K_{cini}.

A reliable K_c is fundamental for applying ET equations in a rain garden. To calculate K_c, LAI is measured regularly from the development stage till the end. To evaluate the two equations, Hess et al., (2019) measured RMSE, R^2, Nash–Sutcliffe coefficient of efficiency and the differences between the observed and predicted data. They optimized each parameter for the intended process and crop coefficient. Hess et al., (2019) also studied three different lysimeters, in a sandy loam with unconstricted outflow (UO), UO sand and sand with internal water storage (IWS). HG produced high ETs in the UO sands and low rates for the IWS system. FAO56PM showed low values with IWS and good values in UO systems. Both equations underestimated the IWS system in the year with heavy spring rain compared with the successive year, with less precipitation. Both equations were based on a constant water volume, and the differences in precipitation explain the errors in the ET equations.

Hess et al., (2019) used SMEF equations when soil moisture was below field capacity and calculated Kc for each equation, and all showed good performance with all three lysimeters. In particular, FAO56PM showed high daily results for the three lysimeters, in particular for the IWS system. HG showed low values for all three systems. The SMEF equations well represented the daily calibrations for the UO lysimeters. For storm scale, the results from both unmodified equations showed very good regression for all three lysimeters. The nonlinear SMEF used for the daily calibration accurately described the UO systems for storm scale. The modified equations showed good estimated daily ET rates, although the authors recommended modifications for storm scale to yield better results for both equations. Hess et al., (2019) calibrated FAO56PM and FG, with and without modification, on the daily data and storm data, and both modified equations estimated relatively accurate ETs.

Weiss et al., (2021) studied how ET behaves in a technical soil, that is, a soil in bioswales and rain gardens; they found large impacts of ET that even affected urban climates. The authors developed an equation for technical soils with and without vegetation and compared it with Makk, which among the radiation-based equations is reliable in very dry conditions. They developed a different equation for each soil,

 DOI: 10.1201/9781003467229-12

with each equation is affected by climate. Weiss et al. (2021)'s equation considered only one parameter, average monthly temperature, whereas Makk considers more parameters (Table 12.1).

The developed equation was based on second-degree polynomial regression:

$$y = \frac{ax^2 + bx + c}{d} \tag{247}$$

y = representative ET [mm/day]; x = the chosen temperature [°C]; a,b,c = constant; d = surface area of the used containers to determine ET of a technical soil per m².

Weiss et al., (2021) tested their equation in six technical soils: soils A and B were applied in a rain garden planted with perennial plants; C was a standard technical soil for planting trees on the streets in Austria; soils D and E were developed for planting trees in better living conditions; and F was a technical soil for rain gardens. C–F were developed to increase the water infiltration during storms in Austria. Table 12.2 gives the technical data on the six soils Weiss et al. studied.

E1 (unplanted technical soil A):

$$y = \frac{0.00004x^2 + 0.0009x + 0.2032}{0.0593957} \tag{248}$$

E2 (planted soil A with *Sedum floriferum*):

$$y = \frac{-0.0002x^2 + 0.01x + 0.0765}{0.0593957} \tag{249}$$

E3 (planted soil A with *Geranium x cantabrigiense*):

$$y = \frac{-0.0002x^2 + 0.0153x + 0.0126}{0.0593957} \tag{250}$$

TABLE 12.1
Parameters for the Makkink Equation and Weiss et al. (2021)'s Equation for Technical Soils

Makkink equation	Technical soil equation
Average monthly air temperature	Average monthly air temperature
Extraterrestrial radiation	
Average monthly sunshine duration	
Monthly wind velocity	
Relative air humidity	
Maximum sunshine duration	

Source: Modified from Weiss et al. (2021).

E4 (unplanted technical soil B):

$$y = \frac{0.0002x^2 + 0.0037x + 0.151}{0.0593957} \tag{251}$$

E5 (planted soil A with *Sedum floriferum*):

$$y = \frac{-0.00001x^2 + 0.0056x + 0.1037}{0.0593957} \tag{252}$$

E6 (planted soil B with *Geranium x cantabrigiense*):

$$y = \frac{0.0001x^2 + 0.0028x + 0.1288}{0.0593957} \tag{253}$$

E7 (unplanted technical soil C):

$$y = \frac{-0.00007x^2 + 0.0114x + 0.0903}{0.0593957} \tag{254}$$

E8 (unplanted technical soil D):

$$y = \frac{0.00002x^2 + 0.0061x + 0.0755}{0.0593957} \tag{255}$$

E9 (unplanted technical soil E):

$$y = \frac{0.0001x^2 - 0.0025x + 0.1537}{0.0593957} \tag{256}$$

TABLE 12.2
ET Data for Technical Soils A–F Planted and Nonplanted

Substrate	ETP [mm/day]	ETP [mm/month]
A-E1	3.43	102.92
A-E2	3.15	94.50
A-E3	4.55	136.42
B-E4	4.31	129.30
B-E5	4.00	119.93
B-E6	4.40	131.98
C-E7	5.58	167.46
D-E8	4.05	121.47
E-E9	2.59	77.63
F-E10	4.55	136.58

Source: Modified from Weiss et al. (2021).

E10 (unplanted technical soil F):

$$y = \frac{-0.0003x^2 - 0.0191x - 0.0196}{0.0593957} \tag{257}$$

The different results from the 10 equations are because it was difficult to transform equations developed from natural soils for application in technical soils, and accuracy was lost. These equations have the advantage of being more precise in technical soils; moreover, they are easy to use.

REFERENCES

Hargreaves, G.H., and Samani, Z.A. Reference crop evapotranspiration from temperature. Appl. Eng. Agric., 1, no 2, (1985): 96–99.

Hess, A., Wadzuk, B., and Welker, A. Predictive evapotranspiration equations in rain gardens. J. Irrig. Drain. Eng., 145, no 7, (2019): 04019010.

Weiss, O., Minixhofer, P., Scharf, B., and Pitha, U. Equation for calculating evapotranspiration of technical soils for urban planting. Land, 10, no 6, (2021): 622.

13 Future Strategies for Reducing ET in a Changing Climate

Temperature, humidity and the stomatal resistance of plants will all be affected by climate change and therefore also by ET_0. Indeed, climate change affects the surface water balance (runoff, soil moisture, groundwater, etc.), drought phenomena, irrigation and agricultural activity and, influencing ET_0, food and water policies. Deep understanding of evapotranspiration will be fundamental for mitigating, abating and even preventing climate change consequences as well as generally being helpful in weather forecasting and managing hydrologic cycles. Understanding ET is critical for the sustainable use of water resources, especially in agricultural activities, in particular given the increasing populations worldwide. However, estimating ET is a long process that is not always effective.

Dadaser-Celik et al., (2016) and Song et al., (2010) determined that net radiation and wind speed had more significant effects on ET_0 than did maximum and minimum air temperature. On the contrary, Darshana et al., (2013) found strong impacts of maximum temperature and net radiation on ET_0, Zhang et al., (2019) described that ET_0 was controlled by maximum air temperature, RH and wind speed.

Meanwhile, Valipour et al., (2020) highlighted the importance of the period, describing that it is important to consider data for a 50-year period since data from shorter periods are more likely measuring climate variability and not climate change. They used 12 meteorological variables for a 50-year period considering areas with arid, semiarid, Mediterranean and humid climates through FAO56MP, Mann–Kendall (TFPW-MK) and Spearman's rho. MK is suitable for inferring trends in a series as a nonparametric statistical test for non-Gaussian data distribution and for showing the specific characteristics in a hydrologic time series (Bakhtiari et al., 2011; Valipour et al., 2020).

Sarr et al., (2015), Salas et al., (1980) and Valipour et al., (2020) identified that hydrologic time series can show correlations that could invalidate the accuracy of trends and that trend-free pre-whitening removes this effect in the MK equation. Spearman's rho calculates the correlations between ET_0 from FAO56 PM and the meteorological variables. Valipour et al., (2020) used Spearman's rho to conclude that wind speed showed the same trends in ET_0 results in many province of Iran. They identified minimum and maximum temperature and $e_s - e_a$ (vapor pressure deficit) as the driving forces in ET_0 equations.

Valipour et al., (2020) found that parameters like mean temperature, minimum RH and wind speed were fundamental, seemingly reflecting that climate change more deeply affects arid than semiarid areas. Indeed, regions worldwide experienced

 DOI: 10.1201/9781003467229-13

increasing maximum, mean and minimum temperature and vapor pressure in arid areas during the 50-year period, whereas in semiarid areas minimum and mean temperatures increased, and minimum RH and cloudy days showed decreasing values. These results confirmed how ET_0 is mainly influenced by both air temperature and wind speed; in particular, the cold seasons affect ET_0 more than the warm seasons. Moreover, anthropogenic activity like land use changes, urban development and desertification affect wind speed and direction, which influences ET_0 (Azami et al., 2015; Khaledian et al., 2012; Tabari et al., 2014; Valipour et al., 2020). The authors concluded, from the FAO56PM, results that ET_0 is influenced by air temperature, air humidity, wind speed and solar radiation but that it ignores stomatal resistance. Stomatal resistance depends on the CO_2 amount and surface albedo, which depends on moisture changes in vegetation and soil. ET_0 is also influenced by vapor pressure deficit, which increases because of the atmospheric request for water vapor. Furthermore, higher temperatures increase the water vapor from soil moisture, which opens a plant's stomata and in turn releases more water vapor.

Land ET is a crucial component of global water and energy cycles; about 60% of precipitation returns to the atmosphere, and almost 50% of solar radiation is absorbed. Climate change and land use changes affect ET and CO_2 concentrations in the atmosphere (Mao et al., 2015; Piao et al., 2007; Shi et al., 2011). Land use change has particularly affected ET in recent decades (Liu et al., 2021).

ET is involved in global water and energy cycles, and its changes are affecting runoff, temperature and all the parameters involved in the interaction between land and atmosphere (Shukla and Mintz, 1982; Liu et al., 2021). It is altering water resources and spatio-temporal variations in regional and global temperatures. In particular, water cycle intensified over the 20th century and is accelerating in the 21st because of climate change (Pan et al., 2015).

This intensification is resulting in increasing drought and flood events; dry areas are becoming drier, and wet areas are getting wetter. Climate change has generally had profound effects on precipitation and PET, that is water and energy (Troch et al., 2013; Liu et al., 2021). High atmospheric CO_2 concentrations reduce stomatal conductance and water transpiration through leaves (Zeng et al., 2018; Liu et al., 2021) and these factors may control ET at regional and seasonal levels (Douville et al., 2013; Mao et al., 2015; Zeng et al., 2018; Liu et al., 2021).

Solar radiation, wind speed and temperature affect ET, controlling water vapor and moisture. Air temperature influences potential evapotranspiration, controlling the air moisture-holding capacity and the water flux from soil and atmosphere. Precipitation influences the soil water content and soil moisture; in fact, in areas where precipitation is the major parameter affecting soil water content, actual evapotranspiration is dominated by precipitation. In areas with low precipitation, ET shows a negative correlation with temperature increase and vice versa, reflected in decreasing soil moisture like in South America, Africa and Australia or in cloudiness caused by decreasing diurnal temperature ranges (Pan et al., 2015). Climate change is causing major changes in precipitation and terrestrial ET.

The Intergovernmental Panel on Climate Change (IPCC 2017) determined that global temperatures have increased by 0.74 °C and will have increased by 4.6 °C by 2090; CO_2 will increase from 379.6 to 809 ppm (scenario A2), deeply affecting ET.

Furthermore, 91.3% of climate variability might influence inter-annual variations in ET (Pan et al., 2015). Increasing CO_2 concentration stimulates plant growth and in turn leaf area and leaf transpiration (Cao et al., 2010; Felzer et al., 2009; Piao et al., 2007, Pan et al., 2015); however, high CO_2 concentrations lead plants to close their stomata and reduce transpiration (Baker et al., 1990; Shams et al., 2012; Pan et al., 2015). Pan et al., (2015) calculated ET using FAO56PM for canopy conductance and resistance:

$$LAI_j^{sun} = 2cos\theta_{ave}\left(1-e^{-0.5\Omega LAI/cos\theta_{ave}}\right) \tag{258}$$

$$LAI^{sha} = LAI - LAI_j^{sun} \tag{259}$$

$$(g_c)j = \frac{LAI_j^{sun}}{(r_s^{sun})_j} + \frac{LAI_j^{sha}}{(r_s^{sha})_j} \tag{260}$$

g_c = canopy conductance [m/s]; r_s^{sun}, r_s^{sha} = stomatal resistance of sunlit leaves and shaded leaves [s/m]; LAI = leaf area index; LAI_j^{sun}, LAI_j^{sha} = leaf area index for sunlit leaves and shaded leaves; Ω = PFT, specific parameter to represent foliage clumping effect; θ_{ave} = solar zenith angle.

Stomatal conductance is a function of photosynthetically active radiation, atmospheric CO_2, maximum and minimum stomatal conductance and vapor pressure deficit:

$$g_s = max(g_{max}r_{corr}bf\ (ppdf)\ f(T_{min})\ f(vpd)\ f(CO_2), g_{min}) \tag{261}$$

$$r_{corr} = \left(\frac{T_{day} + 273.15}{293.15}\right)^{1.75}\left(\frac{101300}{p}\right) \tag{262}$$

$$f\ (ppdf) = \frac{ppdf}{75 + ppdf} \tag{263}$$

$$b_i = 1 \longrightarrow \psi_i > \psi_{open} \tag{264}$$

$$\frac{\psi_{open} - \psi}{\psi_{open} - \psi_{close}}\psi_{close} \le \psi_i \le \psi_{open} \tag{265}$$

$$0 \longrightarrow \psi_i < \psi_{close} \tag{266}$$

$$b = \sum_{i=1}^{10}\left(root_i\frac{\theta_{sat,i}-\theta_{ice,i}}{\theta_{sat,i}}b_i\right) \tag{267}$$

$$f\ (T_{min}) = 1\ Tmin > 0\ °C$$

$$1 + 0.125 T_{min} - 8\ °C \le T_{min} \le 0° \tag{268}$$

$$0\ T_{min} < -8\ °C$$

$$f\,(vpd) = 1\ vpd < vpd_{open}$$

$$\frac{vpd_{close} - vpd}{vpd_{close} - vpd_{open}}\ vpd_{open} \le vpd \le vpd_{close} \tag{269}$$

$$0\ vpd > vpd_{close}$$

$$f\,(CO_2) = -0.001 CO_2 + 1.35 \tag{270}$$

gs = stomatal conductance; rcorr = correction factor of temperature and air pressure on conductance; b = soil moisture factor; ppdf = photosynthetic photo flux density [(umol/m^{2}s^1]; Tmin = daily minimum T [°C]; vpd = vapor pressure deficit [pa]; CO_2 = atmospheric CO_2 concentration [ppm]; gmax, gmin = maximum and minimum stomatal conductance for vapor [m/s]; p = air pressure [Pa]; θsat,i = ith soil layer saturated volumetric water content [mm H_2O/m^2]; θice,i = volumetric ice content of the ith soil layer [mm H_2O/m^2]; Ψ = water potential [mm H_2O]; Ψopen, Ψclose = water potential under which the stomata fully open and close [mm H_2O]; vpdclose, vpdopen = vapor pressure deficit (vpd) when the leaf stomata is fully closed and open, respectively [Pa].

Pan et al., (2015) found that on irrigated cropland, water transpiration deficits can be replenished by irrigation. The canopy influences net radiation and aerodynamic resistance, and measuring potential soil evapotranspiration (PSE) is based on short-wave radiation crossing the canopy as an energy source. PSE is corrected according to the leaf area:

$$EVAP = pet_{PM}e^{-0.6LAI} \tag{271}$$

EVAP = soil evaporation; pet$_{PM}$ = PET estimated with FAO56PM; LAI = average LAI over the land area in each grid.

Pan et al., (2015) calculated the effects of β to determine the CO_2 effect on ET:

$$ET_{CO2} = \frac{ET_{clm+CO_2} - ET_{clm}}{CO_{2concentration}\,(ppm)} \tag{272}$$

In the dynamic land ecosystem model, the authors investigated the considered dataset to measure terrestrial ET from 2000 to 2009 on the following parameters: daily climate, annual land use and land cover, nitrogen, ozone, carbon dioxide, irrigation, fertilizer use, river networks, cropping system information and soil properties.

Pan et al., (2015) considered two projections from the IPCC: scenarios A2 and B1. The former is based on an increasing population and economic development; the latter is less material intensive, with clean, resource-efficient technology being developed over the period from 2010 to 2099 considering nitrogen deposition, ozone pollution and land use constant from 2010. The IPCC modeled that temperature will increase by 4.6 °C and 1.5 °C, respectively in scenarios A2 and B1 by the year 2090 compared with 2000 values; precipitation will increase by 16.9% and 7.5% and CO_2 concentration will increase by 114.9% and 44.5%.

According to the model, in 2090, terrestrial ET will be 14% and 4.5% higher, respectively for scenarios A2 and B1; these were the only two scenarios showing this increase during the 21st century. Terrestrial ET increases more in A2 than in B1 because of the increase of the air temperature. Indeed, terrestrial ET will rise by 9.5 mm and 28 mm in A2 and B1 from 2010 to 2099, while at the same time, precipitation will rise without deeply affecting the two scenarios. In contrast, climate change and CO_2 concentration will cause terrestrial ET to increase in B1 from 2000 to 2099 with low correlation with climate factors; meanwhile, the increasing ET in A2 is mainly caused by the air temperature (Figure 13.1a) and precipitation (Figure 13.1b) (Pan et al., 2015).

Climate change and high CO_2 concentration will cause low increases in terrestrial ET than will climate change alone in both scenarios (Figure 13.2); throughout the 21st century, there will be low increases in global terrestrial ET, mainly in A2 (5.6%). B1 with CO_2 simulation show a very low increase in terrestrial ET (2.4%) by 2090 compared with data from 2000 and faint correlations between ET and climate factors. A2 shows good correlations with increasing precipitation and surface air temperature (Pan et al., 2015).

Terrestrial ET will vary according to the latitude because of climate change during the 21st century (see Table 13.1); the changes will be high at low latitudes and low at high latitudes. A2 shows increases in terrestrial ET of 21.7% at high latitude and 13.3% at low latitude. In B1, however, terrestrial ET decreases after 2060 at high latitude and increases continuously at middle and low latitudes from 2010 to 2099. Terrestrial ET increases by 25–75 mm/yr in A2 in boreal and Arctic regions after 2090; in South America and Africa, the increase in A2 will be 225–275 mm/yr. On the contrary, B1 shows increases of 25–75 mm/yr with the exception of Australia and Africa, with increases over 75 mm/yr. Australia is an exception because in both scenarios it is observed a natural decreasing in terrestrial ET due to the less availability of moisture because of higher temperature (Pan et al., 2015).

Adding high CO_2 concentration and climate warming to A2, ET increased by 15.5% at the mid-latitudes by the end of the 21st century; in B1, there is a low increase 4.5% at high latitudes by the 2090s. In A2, terrestrial ET decreases in areas at low latitudes (South America, Central Africa and Southeast Asia); B1 shows lower decreases of 30–50 mm/yr in South America and Central Africa (Pan et al., 2015).

Considering the β effect to calculate the consequences of CO_2 concentration on terrestrial ET, high CO_2 lowers ET in both scenarios, more in A2 than in B1. Pan et al., (2015) calculated that CO_2 fertilization would save 56.8 μmm^2/yr of water in A2 but less in B1 because of less atmospheric CO_2. The β effect will be lowest in areas at high latitudes in A2, but in B1, the fertilization reduces ET. High atmospheric CO_2 amount will also reduce stomatal aperture, with reductions in canopy ET and plant transpiration, although the latter consequence could be mitigated by increasing the LAI. Gedney et al., (2006) found that for the same reasons, runoff will increase at continental scale. Table 13.2 displays Pan et al.'s (2015) calculations

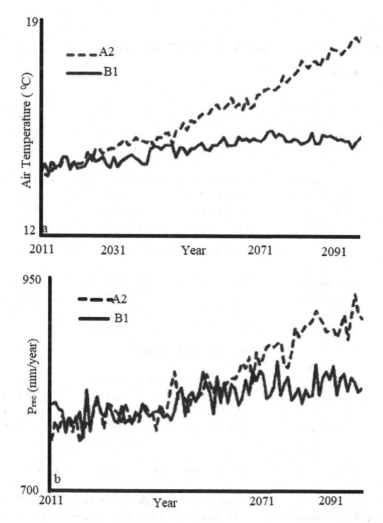

FIGURE 13.1 Temporal pattern analysis from 2010 to 2099: (a) temperature; (b) precipitation for scenarios A2 and B1.

Source: Modified from Pan et al. (2015).

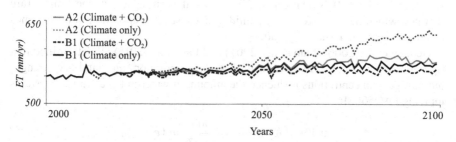

FIGURE 13.2 Global changes in ET according to climate and CO_2, IPCC scenarios A2 and B1.

Source: Modified from Pan et al. (2015).

TABLE 13.1
Global ET by Latitude under the Four IPCC Scenarios

Global ET (mm/yr)

	A2 (Climate only)	A2 (Climate + CO_2)	B1 (Climate only)	B1 (Climate + CO_2)
2000s	549.25			
2030s	564.16(2.7%)	558.96(1.8%)	559.41(1.9%)	555.54(1.1%)
2060s	590.93(7.6%)	573.28(4.4%)	571.84(4.1%)	562.71(2.5%)
2090s	626.15(14.0%)	579.96(5.6%)	573.87(4.5%)	562.40(2.4%)
2000s	781.97			
2030s	804.60(2.9%)	796.90(1.9%)	799.74(2.3%)	793.94(1.5%)
2060s	843.45(7.9%)	817.02(4.5%)	813.35(4.0%)	799.75(2.3%)
2090s	886.02(13.3%)	816.75(4.4%)	817.46(4.5%)	800.54(2.4%)

Mid-Latitude ET (mm/yr)

	A2 (Climate only)	A2 (Climate + CO_2)	B1 (Climate only)	B1 (Climate + CO_2)
2000s	366.64			
2030s	375.55(2.4%)	372.19(1.5%)	368.42(0.5%)	365.97(−0.2%)
2060s	391.67(6.8%)	380.71(3.8%)	381.46(4.0%)	375.60(2.4%)
2090s	419.35(14.4%)	391.38(6.7%)	381.63(4.1%)	374.11(2.0%)

High Latitude ET (mm/yr)

	A2 (Climate only)	A2 (Climate + CO_2)	B1 (Climate only)	B1 (Climate + CO_2)
2000s	212.87			
2030s	216.50(1.7%)	215.28(1.1%)	219.18(3.0%)	218.37(2.6%)
2060s	229.23(7.7%)	224.83(5.6)	225.36(5.9%)	223.38(4.9%)
2090s	259.04(21.7%)	245.91(15.5%)	225.00(5.7%)	222.40(4.5%)

Source: Modified from Pan et al. (2015).

of how ET changed for a number of plants in the decade from 2000 to 2009. Figure 13.3 graphically presents how CO_2 fertilization effects will vary by latitude in both scenarios.

Swelam et al. (2010) found that climate change is causing increased air temperature, humidity and CO_2 and that temperature can reduce these impacts on ET (Chaouche et al., 2010; Swelam et al., 2010). Global warming affects the temperature of water, which in turn affects the humidity because its evaporation into the atmosphere increases with the temperature.

Gedney et al. (2006), Piao et al. (2007) and Pan et al. (2015) all determined that high atmospheric CO_2 affects terrestrial water evaporation, and land use changes and nitrogen concentrations influence the amount of water in terrestrial ecosystems applying FAO56PM:

$$ET_0 = \frac{0.408\Delta\left(R_n - G\right) + \gamma \frac{900}{T+273} u_2 \left(e_s - e_a\right)}{\Delta + \gamma\left(1 + 0.34u_2\right)}$$

(12)

TABLE 13.2
Mean Decadal ET and ET Changes in Plants for the Decade 2000–2009 under Each IPCC Scenario

			Tundra	BNEF	BNDF	TBDF	TBEF	TNEF	TrBDF	TrBEF	Deciduos shrub	Evergreen shrub	C3 grass	Herbaceous wetland	Woody wetland	Cropland
B2	Net Change	Climate +CO$_2$	8.6	11.1	27.5	9.1	5.4	10.0	24.6	24.3	6.9	19.1	15.7	12.4	4.2	14.5
		Climate only	10.8	15.2	33.5	28.9	34.5	20.4	48.3	62.4	12.7	29.0	20.8	19.6	28.8	26.3
A2		Climate + CO$_2$	42.6	27.9	75.3	26.9	45.9	35.6	-3.9	75.4	24.3	15.5	27.8	21.3	7.7	38.9
		Climate only	53.2	45.4	101.1	98.0	154.7	75.1	89.2	224.3	49.2	61.7	50.0	55.3	108.4	83.7
Decadal 2000s Mean		(mm/yr)	200.39	316.57	309.13	598.79	981.28	443.55	1000.41	1465.27	313.23	620.10	358.61	314.76	987.59	620.87

Source: Modified from Pan et al. (2015).

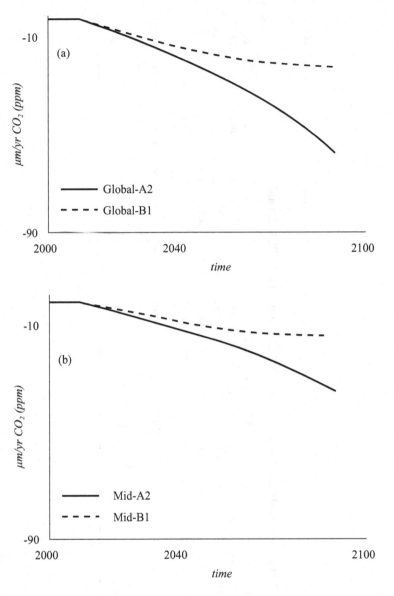

FIGURE 13.3 CO_2 fertilization effects by latitude: (a) global; (b) mid-latitude.

Source: Modified from Pan et al. (2015).

Considering the projected CO_2 amount of 550 ppm for 2050 and air temperature over 3.0 °C, this scenario will be the worst in California according to projections from Prentice et al., (2001) and Cayan et al., (2006).

Swelam et al., (2010), in a test in which they increased the dew point temperature and canopy resistance, observed a compensation effect of temperature on ET_0. They determined that air temperature increase of 4 °C with canopy resistance remaining at 70 s/m will be the worst case. ET_0 was higher than the current annual value under low wind speeds and was lower in high winds. Based only on increasing temperatures, these worst-case scenarios for decreasing ET are plausible as water resources are depleted unless stakeholders worldwide undertake good planning.

Estimating the necessary water amounts for agricultural activity requires intensive planning and management; sustainable use of water is necessary to mitigate the impacts of its scarcity and waste. Increasing the efficiency of water management, by better estimating ET, is a robust way of planning appropriately for future water use. Climate change and global warming affect ET and agriculture activity. Climate change causes drought and floods that affect the amount of water in soil; warming increases temperature and affects precipitation, which both affect agriculture production. In fact, according to Wanniarachchi and Sarukkalige, (2022) and Aryalekshmi et al., (2021), about 70% of water consumption is due to irrigation, and it is the main cause of water depletion in many countries; therefore, sustainable management to reduce water waste is fundamental.

The first step is identifying a suitable ET equation for optimizing agricultural water use given climate change (Krishna, 2018; Wanniarachchi and Sarukkalige, 2022). Wanniarachchi and Sarukkalige, (2022) suggested studying the climate change impacts on ET and water availability using regional rather than global climate models, which are too coarse. Incorporating technology into increasing water efficiency and sustainability for the future is mandatory, and remote sensing and satellites have proven immeasurably useful tools for measuring ET in different areas. The standard equation FAO56PM is the most trusted, but it is difficult to compute when meteorological data are not available.

Lang et al., (2017) determined that radiation-based methods better measure PET than temperature-based methods after comparing the radiation-based methods Makkink, Abtew and Priestley–Taylor and the temperature-based methods Hargreaves–Samani, Thornthwaite, Hamon, Linacre and Blaney–Criddle with FAO56PM. In particular, Makk and Abtew are suitable for warm areas at low latitudes, and Makk is the most suitable in areas with complex geographic features. Remote sensing is another tool for calculating ET and thereby the necessary amount of water for agriculture activity. Reyes-Gonzáles et al., (2018) highlighted the use of remote sensing to calculate the necessary crop water for regional and field scale, and found remote sensing to be potentially ideal when meteorological data are scarce. Remote sensing uses land surface temperature, and it allows for determining potential crop transpiration using the crop coefficient and actual evaporation.

REFERENCES

Aryalekshmi, B., Biradar, R.C., Chandrasekar, K., and Ahamed, J.M. Analysis of various surface energy balance models for evapotranspiration estimation using satellite data. Egypt. J. Remote Sens. Space Sci., 24, (2021): 1119–1126.

Azami, M., Mirzaee, E., and Mohammadi, A. Recognition of urban unsustainability in Iran (case study: Sanandaj City). Cities, 49, (2015): 159–168.

Baker, J., Allen, L., Boote, K., Jones, P., and Jones, J. Rice photosynthesis and evapotranspiration in subambient, ambient, and superambient carbon dioxide concentrations. Agron. J., 82, no 4, (1990): 834–840.

Bakhtiari, B., Ghahraman, N., Liaghat, A.M., and Hoogenboom, G. Evaluation of reference evapotranspiration models for a semiarid environment using lysimeter measurements. J. Agric. Sci. Technol., 13, (2011): 223–237.

Cao, L., Bala, G., Caldeira, K., Nemani, R., and Ban-Weiss, G. Importance of carbon dioxide physiological forcing to future climate change. Proc. Natl. Acad. Sci. U.S.A., 107, no 21, (2010): 9513–9518.

Cayan, D., Luers, A.L., Hanemann, M., and Franco, G. Scenarios of climate change in California: An overview. California Energy Commission, CEC-500–2005–186-SF. (2006).

Chaouche, K., Neppel, L., Dieulin, C., Pujol, N., Ladouche, B., Martin, E., Salas, D., and Caballero, Y. Analyses of precipitation, temperature and evapotranspiration in a French Mediterranean region in the context of climate change. Compt. Rendus Geosci., 342, no 3, (2010): 234–243.

Dadaser-Celik, F., Cengiz, E., and Guzel, O. Trends in reference evapotranspiration in Turkey: 1975–2006. Int. J. Climatol., 36, (2016): 1733–1743.

Darshana, P.A., and Pandey, R.P. Analysing trends in reference evapotranspiration and weather variables in the Tons River Basin in Central India. Stoch. Environ. Res. Risk Assess., 27, (2013): 1407–1421.

Douville, H., Ribes, A., Decharme, B., Alkama, R., and Sheffield, J. Anthropogenic influence on multidecadal changes in reconstructed global evapotranspiration. Nat. Clim. Chang., 3, no 1, (2013): 59–62.

Felzer, B.S., Cronin, T.W., Melillo, J.M., Kicklighter, D.W., and Schlosser, C.A. Importance of carbon-nitrogen interactions and ozone on ecosystem hydrology during the 21st century. J. Geophys. Res., 114, no G1, (2009): 1–10.

Gedney, N., Cox, P.M., Betts, R.A., Boucher, O., Huntingford, C., and Stott, P.A. Detection of a direct carbon dioxide effect in continental river runoff records. Nature, 439, no 7078, (2006): 835–838.

Khaledian, Y., Kiani, F., and Ebrahimi, S. The effect of land use change on soil and water quality in northern Iran. J. Mount. Sci., 9, (2012): 798–816.

Krishna, P.A. Evapotranspiration and agriculture: A review. Agric. Rev., 40, (2018): 1–11.

Lang, D., Zheng, J., Shi, J., Liao, F., Ma, X., Wang, W., Chen, X., and Zhang, M. A comparative study of potential evapotranspiration estimation by eight methods with FAO Penman—Monteith method in southwestern China. Water, 9, no 10, (2017): 734–752.

Liu, J., You, Y., Li, J., Sitch, S., Gu, X., Nabel, J.E., Lombardozzi, D., Luo, M., Feng, X., Almut, A., Jain, A.K., Friedlingstein, P., Tian, H., Poulter, B., and Kong, D. Response of global land evapotranspiration to climate change, elevated CO2, and land use change. Agr. For. Meteorol., 311, (2021): 108663.

Mao, J.F., Fu, W.T., Shi, X.Y., Ricciuto, D.M., Fisher, J.B., Dickinson, R.E., Wei, Y.X., Shem, W., Piao, S.L., Wang, K.C., Schwalm, C.R., Tian, H., Mu, M., Arain, A., Ciais, P., Cook, R., Dai, Y., Hayes, D., Hoffman, F.M., Huang, M., Huang, S., Huntzinger, D.N., Ito, A., Jain, A., King, A.W., Lei, H., Lu, C., Michalak, A.M., Parazoo, N., Peng, C., Peng, S., Poulter, B., Schaefer, K., Jafarov, E., Thornton, P.E., Wang, W., Zeng, N., Zeng, Z., Zhao, F., Zhu, Q., and Zhu, Z. Disentangling climatic and anthropogenic controls on global terrestrial evapotranspiration trends. Environ. Res. Lett., 10, no 9, (2015): 1–13.

Pan, S., Tian, H., Dangal, S.R., Yang, Q., Yang, J., Lu, C., and Ouyang, Z. Responses of global terrestrial evapotranspiration to climate change and increasing atmospheric CO_2 in the 21st century. Earth's Future, 3, no 1, (2015): 15–35.

Piao, S., Friedlingstein, P., Ciais, P., de Noblet-Ducoudre, N., Labat, D., and Zaehle, S. Changes in climate and land use have a larger direct impact than rising CO2 on global river runoff trends. Proc. Natl. Acad. Sci., 104, no 39, (2007): 15242–15247.

Prentice, I.C., Farquhar, G.D., Fasham, M.J.R., Goulden, M.L., Heimann, M., Jaramillo, V.J., Kheshgi, H.S., Le Quéré, C., Scholes, R.J., and Wallace, D.W.R. The carbon cycle and atmospheric carbon dioxide. In: Climate Change 2001: The Scientific Basis: Contribution of Working Group I to the Third Assessment Report of the Intergovernmental Panel on Climate Change. Cambridge University, Cambridge, United Kingdom and New York, NY, USA, (2001): 183–238.

Reyes-González, A., Kjaersgaard, J., Trooien, T., Hay, C., and Ahiablame, L. Estimation of crop evapotranspiration using satellite remote sensing-based vegetation index. Adv. Meteorol., (2018): 1–12.

Salas, J.D., Delleur, J.W., Yevjevich, V., and Lane, W.L. Applied Modeling of Hydrologic Time Series. Water Resources Publication, Littleton, CO, USA, (1980).

Sarr, M.A., Gachon, P., Seidou, O., Bryant, C.R., Ndione, J.A., and Comby, J. Inconsistent linear trends in Senegalese rainfall indices from 1950 to 2007. Hydrol. Sci. J., 60, (2015): 1538–1549.

Shams, S., Nazemosadat, S., Haghighi, A.K., and Parsa, S.Z. Effect of carbon dioxide concentration and irrigation level on evapotranspiration and yield of red bean. J. Sci. Technol. Greenhouse Cult., 2, no 8, (2012): 1–10.

Shi, X., Mao, J., Thornton, P.E., Hoffman, F.M., and Post, W.M. The impact of climate, CO2, nitrogen deposition and land use change on simulated contemporary global river flow. Geophys. Res. Lett., 38, no 8, (2011): 1–6.

Shukla, J., and Mintz, Y. Influence of land-surface evapo-transpiration on the Earth's climate. Science, 215, no 4539, (1982): 1498–1501.

Song, Z.W., Zhang, H.L., Snyder, R.L., Anderson, F.E., and Chen, F. Distribution and trends in reference evapotranspiration in the North China: Plain. J. Irrig. Drain. Eng., 136, (2010): 240–247.

Swelam, A., Jomaa, I., Shapland, T., Snyder, R.L., and Moratiel, R. Evapotranspiration response to climate change. In XXVIII International Horticultural Congress on Science and Horticulture for People (IHC2010), International Symposium on 922. (2010): 91–98.

Tabari, H., Talaee, P.H., Nadoushani, S.M., Willems, P., and Marchetto, A. A survey of temperature and precipitation based aridity indices in Iran. Quater. Int., 345, (2014): 158–166.

Troch, P.A., Carrillo, G., Sivapalan, M., Wagener, T., and Sawicz, K. Climate-vegetation-soil interactions and long-term hydrologic partitioning: Signatures of catchment co-evolution. Hydrol. Earth Syst. Sci., 17, no 6, (2013): 2209–2217.

Valipour, M., Bateni, S.M., Gholami Sefidkouhi, M.A., Raeini-Sarjaz, M., and Singh, V.P. Complexity of forces driving trend of reference evapotranspiration and signals of climate change. Atmosphere, 11, no 10, (2020): 1081.

Wanniarachchi, S., and Sarukkalige, R. A review on evapotranspiration estimation in agricultural water management: Past, present, and future. Hydrol., 9, no 7, (2022): 123.

Zeng, Z.Z., Peng, L.Q., and Piao, S.L. Response of terrestrial evapotranspiration to Earth's greening. Curr. Opin. Environ. Sustain., 33, (2018): 9–25.

Zhang, L., Traore, S., Cui, Y., Luo, Y., Zhu, G., Liu, B., Fipps, G., Karthikeyan, R., and Singh, V. Assessment of spatiotemporal variability of reference evapotranspiration and controlling climate factors over decades in China using geospatial techniques. Agric. Water Manag., 213, (2019): 499–511.

References

Abedi-Koupai, J., Amiri, M.J., and Eslamian, S.S. Comparison of artificial neural network and physically based models for estimating of reference evapotranspiration in greenhouse. Aust. J. Basic Appl. Sci., 3, no 3, (2009): 2528–2535.

Abtew, W. Evapotranspiration measurements and modeling for three wetland systems in south Florida. J. Am. Water Resour. Assoc., 32, no 3, (1996): 465–473.

Adeboye, O.B., Osunbitan, J.A., Adekalu, K.O., and Okunade, D.A. Evaluation of FAO-56 penman-monteith and temperature based models in estimating reference evapotranspiration using complete and limited Data, application to Nigeria. Agric. Eng. Int.: CIGR J., 11, (2009): 1–25.

Agam, N., Evett, S.R., Tolk, J.A., Kustas, W.P., Colaizzi, P.D., Alfieri, J.G., Mckee, L.G., Copeland, K.S., Howell, T.A., and Chavez, J.L. Evaporative loss from irrigated inter rows in a highly advective semi-arid agricultural area. Adv. Water Res., 50, (2012): 20–30.

Akhavan, S., Kanani, E., and Dehghanisanij, H. Assessment of different reference evapotranspiration models to estimate the actual evapotranspiration of corn (Zea mays L.) in a semiarid region (case study, Karaj, Iran). Theor. Appl. Climatol., 137, (2019): 1403–1419.

Albrecht, F. Die Methoden zur Bestimmung Verdunstung der natürlichen Erdoberfläche. Arc. Meteor. Geoph. Bioklimatol. Ser. B, 2, (1950): 1–38.

Allen, R.G., Morse, A., Tasumi, M., Trezza, R., Bastiaanssen, W.G.M., Wright, J.L., and Kramber, W. Evapotranspiration from a satellite-based surface energy balance for the Snake River Plan aquifer in Idaho. In: Proceeding of the USCID/EWRI Conference on Energy, Climate, Environment, and Water. U.S. Committee on Irrigation and Drainage, Denver, CO, USA. (2002a).

Allen, R.G., Pereira, L.S., Howell, T.A., and Jensen, M.E. Evapotranspiration information reporting: I. Factors governing measurement accuracy. Agric. Water Manag., 98, (2011): 899–920.

Allen, R.G., Pereira, L.S., Raes, D., and Smith, M. Crop evapotranspiration: Guidelines for computing crop water requirements. In: United Nations FAO, Irrigation and Drainage Paper 56. FAO, Rome, Italy. (1998).

Allen, R.G., and Pruitt, W.O. FAO-24 reference evapotranspiration factors. J. Irrig. Drain. Eng., 117, no 5, (1991): 758–773.

Allen, R.G., Smith, M., and Pereira, L.S. An update for the definition of reference evapotranspiration. ICID Bull., 43, (1994): 1–34.

Allen, R.G., Tasumi, M., and Trezza, R. Satellite-based energy balance for Mapping Evapotranspiration with Internalized Calibration (METRIC) model. J. Irrig. Drain. Eng., 133, (2007b): 380–394.

Allen, R.G., Tasumi, M., and Trezza, R. SEBAL (Surface Energy Balance Algorithms for Land), Advanced Training and User's Manual, Idaho Implementation, Version 1.0, USA. (2002b).

Allen, R.G., Wright, J.L., Pruitt, W.O., Pereira, L.S., and Jensen, M.E. Water requirements. In: Hoffman, G.J., Evans, R.G., Jensen, M.E., Martin, D.L., and Elliot, R.L. (eds.). Design and Operation of Farm Irrigation Systems, 2nd ed. ASABE, St. Joseph, MI, USA, (2007a): 208–288.

Allen, S.J. Measurement and estimation of evaporation from soil under sparse barley crops in Northern Syria. Agric. For. Meteorol., 49, (1990): 291–309.

Amiri, M.J., Zarei, A.R., Abedi-Koupai, J., and Eslamian, S.T. The performance of fuzzy regression method for estimating of reference evapotranspiration under controlled environment. Int. J. Hydrol. Sci. Tech., 9, no 1, (2019): 28–38.

Anderson, M.C., Norman, J.M., Diak, G.R., Kustas, W.P., and Mecikalski, J.R. A two-source time-integrated model for estimating surface fluxes using thermal infrared remote sensing. Remote Sens. Environ., 60, (1997): 195–216.

Armanios, S., El Quosy, D., Abdel Monim, Y.K., and Gaafar, I. Evaluation of evapotranspiration equations using weighing Lysimeter data. Sci. Bull., 35, no 1, (2000): 193–205.

Aryalekshmi, B., Biradar, R.C., Chandrasekar, K., and Ahamed, J.M. Analysis of various surface energy balance models for evapotranspiration estimation using satellite data. Egypt. J. Remote Sens. Space Sci., 24, (2021): 1119–1126.

ASCE-EWRI the ASCE standardized reference evapotranspiration equation. In: ASCE-EWRI Standardization of Reference Evapotranspiration Task Committee Rep., ASCE, Reston, VA, USA. (2005).

Ashktorab, H., Pruitt, W.O., Pawu, K.T., and George, W.V. Energy balance determinations close to the soil surface using a micro-Bowen ratio system. Agric. For. Meteorol., 46, (1989): 259–274.

Ashraf, M., Loftis, J.C., and Hubbard, K.G. Application of geostatistics to evaluate partial weather station networks. Agr. For. Meteorol., 84, (1997): 255–271.

Azami, M., Mirzaee, E., and Mohammadi, A. Recognition of urban unsustainability in Iran (case study: Sanandaj City). Cities, 49, (2015): 159–168.

Azhar, A.H., Masood, M., Nabi, G., and Basharat, M. Performance evaluation of reference evapotranspiration equations under semiarid Pakistani conditions. Arab. J. Sci. En., 39, (2014): 5509–5520.

Babovic, V., and Keijzer, M. Genetic programming as model induction engine. J. Hydroinform., 2, no 1, (2000): 35–60.

Baier, W., and Robertson, G.W. Estimation of latent evaporation from simple weather observations. Can. J. Plant Sci., 45, no 3, (1965): 276–284.

Baker, J., Allen, L., Boote, K., Jones, P., and Jones, J. Rice photosynthesis and evapotranspiration in subambient, ambient, and superambient carbon dioxide concentrations. Agron. J., 82, no 4, (1990): 834–840.

Bakhtiari, B., Ghahraman, N., Liaghat, A.M., and Hoogenboom, G. Evaluation of reference evapotranspiration models for a semiarid environment using lysimeter measurements. J. Agric. Sci. Technol., 13, (2011): 223–237.

Baldocchi, D.D., and Meyers, P.T. Trace gas exchange above the floor of a deciduous forest 1. Evaporation and CO_2 efflux. J. Geophys. Res., 96, (1991): 7271–7285.

Bastiaanssen, W.G.M. SEBAL-based sensible and latent heat fluxes in the irrigated Gediz Basin, Turkey. J. Hydrol., 229, (2000): 87–100.

Bastiaanssen, W.G.M., Menenti, M., Feddes, R.A., and Holtslag, A.A.M. A remote sensing Surface Energy Balance Algorithm for Land (SEBAL) 1: Formulation. J. Hydrol., 212–213, (1998a): 198–212.

Bastiaanssen, W.G.M., Pelgrum, H., Wang, J., Ma, Y., Moreno, J.F., Roenrink, G.J., and van der Wal, T. A remote sensing Surface Energy Balance Algorithm for Land (SEBAL) 2., Validation. J. Hydrol., 212–213, (1998b): 213–229.

Bechini, L., Ducco, G., Donatelli, M., and Stein, A. Modelling, interpolation and stochastic simulation in space and time of global solar radiation. Agric. Ecosyst. Environ., 81, (2000): 29–42.

Ben-Asher, J., Matthias, A.D., and Warrick, A.W. Assessment of evaporation from bare soil by infrared thermometry. Soil Sci. Soc. Am. J., 47, (1983): 185–191.

Benli, B., Bruggeman, A., Oweis, T., and Üstün, H. Performance of penman-monteith FAO56 in a semiarid highland environment. J. Irrig. Drain. Eng., 136, no 11, (2010): 757–765.

Benli, B., Kodal, S., Ilbeyi, A., and Ustun, H. Determination of evapotranspiration and basal crop coefficient of alfalfa with a weighing lysimeter. Agric. Water Manag., 81, no 3, (2006): 358–370.

Berengena, J., and Gavilán, P. Reference evapotranspiration estimation in a highly advective semiarid environment. J. Irrig. Drain. Eng., 131, no 2, (2005): 147–163.

Bezerra, B.G., da Silva, B.B., dos Santos, C.A.C., and Bezerra, J.R.C. Actual evapotranspiration estimation using remote sensing: Comparison of SEBAL and SSEB approaches. Adv. Remote Sens., 4, (2015): 234–247.

Bezerra, B.G., Santos, C.A.C., Silva, B.B., Perez-Marin, A.M., Bezerra, M.V.C., Bezerra, J.R.C., and Ramana Rao, T.V. Estimation of soil moisture in the root-zone from remote sensing data. Rev. Bras. de Ciênc. do Solo, 37, (2013): 595–603.

Blaney, H.F., and Criddle, W.D. Determining consumptive use and irrigation water requirements (No. 1275). US Department of Agriculture. (1962).

Blaney, H.F., and Criddle, W.D. Determining water requirements in irrigated areas from climatological and irrigation data. In: Soil Conservation Service Technical Paper 96, US. Department of Agriculture, Washington, USA. (1950).

Boast, C.W., and Robertson, T.M. A micro-lysimeter method for determining evaporation from bare soil: Description and laboratory evaluation. Soil Sci. Soc. Am. J., 46, (1982): 689–696.

Bogawski, P., and Bednorz, E. Comparison and validation of selected evapotranspiration models for conditions in Poland (Central Europe). Water Resour. Manage., 28, (2014): 5021–5038.

Bouwer, L.M., Biggs, T.W., and Aerts, C.J.H. Estimates of spatial variation in evaporation using satellite-derived surface temperature and a water balance model. Hydrol. Process., 22, (2008): 670–682.

Bowen, I.S. The ratio of heat losses by conduction and by evaporation from any water surface. Phys. Rev., 27, no 6, (1926): 779–787.

Brutsaert, W. Evaporation into the Atmosphere: Theory, History and Applications. D. Reidel Publishing, Dordrecht, The Netherlands, (1982).

Brutsaert, W., and Chen, D.Y. Diagnostics of land surface spatial variability and water vapor flux. J. Geophys. Res., Atmospheres, 100, no D12, (1995): 25595–25606.

Brutsaert, W., and Sugita, M. Application of self-preservation in the diurnal evolution of the surface energy budget to determine daily evaporation. J. Geophys. Res., 97, (1992): 377–382.

Burman, R., and Pochop, L.O. Evaporation, evapotranspiration and climatic data. Developments in Atmospheric Science, 22. Elsevier, Amsterdam, (1994).

Cao, L., Bala, G., Caldeira, K., Nemani, R., and Ban-Weiss, G. Importance of carbon dioxide physiological forcing to future climate change. Proc. Natl. Acad. Sci. U.S.A., 107, no 21, (2010): 9513–9518.

Caprio, J.M. The solar thermal unit concept in problems related to plant development and potential evapotranspiration. In: Lieth, H. (ed.). Phenology and Seasonality Modeling, Ecological Studies, Vol. 8. Springer Verlag, New York, USA, (1974): 353–364.

Castañeda, L., and Rao, P. Comparison of methods for estimating reference evapotranspiration in southern California. J. Environ. Hydrol., 13, no 14, (2005): 1–10.

Caruso, C., and Quarta, F. Interpolation methods comparison. Comput. Math. Appl., 35, no 12, (1998): 109–126.

Cayan, D., Luers, A.L., Hanemann, M., and Franco, G. Scenarios of climate change in California: An overview. California Energy Commission, CEC-500-2005-186-SF. (2006).

Chavez, J.L., Neale, C.M.U., Hipps, L.E., Prueger, J.H., and Kustas, W.P. Comparing aircraft-based remotely sensed energy balance fluxes with eddy covariance tower data using heat flux source area functions. J. Hydromteorol., 6, (2005): 923–940.

Choudhury, B.J., Ahmed, N.U., Idso, S.B., Reginato, R.J., and Daughtry, C.S.T. Relations between evaporation coefficients and vegetation indices studied by model simulations. Remote Sens. Environ., 50, (1994): 1–17.

Chowdhury, A., Gupta, D., Das, D.P., and Bhowmick, A. Comparison of different evapotranspiration estimation techniques for Mohanpur, Nadia district, West Bengal. Int. J. Comput. Eng. Res., 7, no 4, (2017): 33–39.

Chuanyan, Z., and Zhongren, N. Estimating water needs of maize (Zea mays L.) using the dual crop coefficient method in the arid region of northwestern China. Afr. J. Agric. Res., 2, no 7, (2007): 325–333.

Crago, R.D. Conservation and variability of the evaporative fraction during the daytime. J. Hydrol., 180, (1996): 173–194.

Christiansen, J.E. Pan evaporation and evapotranspiration from climatic data. J. Irrig. Drai. Div., 94, no 2, (1968): 243–266.

Daamen, C.C., and Simmonds, L.P. Soil Water, Energy and Transpiration—A Numerical Model of Water and Energy Fluxes in Soil Profiles and Sparse Canopies. Department of Soil Science, University of Reading, Reading, UK, (1994).

Dadaser-Celik, F., Cengiz, E., and Guzel, O. Trends in reference evapotranspiration in Turkey: 1975–2006. Int. J. Climatol., 36, (2016): 1733–1743.

Darshana, P.A., and Pandey, R.P. Analysing trends in reference evapotranspiration and weather variables in the Tons River Basin in Central India. Stoch. Environ. Res. Risk Assess., 27, (2013): 1407–1421.

de Bruin, H. The determination of (reference crop) evapotranspiration from routine weather data. In: Proceedings of Technical Meeting 38, Evaporation in Relation to Hydrology. Committee for Hydrological Research TNO, The Hague, Netherlands, 28. (1981): 25–37.

de Bruin, H., and Lablans, W.N. Reference crop evapotranspiration determined with a modified Makkink equation. Hydrol. Process., 12, no 7, (1998): 1053–1062.

DehghaniSanij, H., Yamamotoa, T., and Rasiah, V. Assessment of evapotranspiration estimation models for use in semi-arid Agricultural. Water Manag., 64, no 2, (2004): 91–106.

Deol, P., Heitman, J.L., Amoozegar, A., Ren, T., and Horton, R. Quantifying nonisothermal subsurface soil water evaporation. Water Resour. Res., 48, (2012): 1–11.

Doorenbos, J., and Pruitt, W.O. Crop Water Requirements. FAO Irrigation and Drainage. Paper No.24. (rev.). FAO, Rome, (1977a).

Doorenbos, J., and Pruitt, W.O. Guidelines for predicting crop water requirements. In: FAO, UN, Irrigation and Drainage Paper No.24., 2nd ed., FAO, Rome, Italy. (1977b).

Douglas, E.M., Jacobs, J.M., Sumner, D.M., and Ram, L.R. A comparison of models for estimating potential evapotranspiration for Florida land cover types. J. Hydrol., 373, (2009): 366–376.

Douville, H., Ribes, A., Decharme, B., Alkama, R., and Sheffield, J. Anthropogenic influence on multidecadal changes in reconstructed global evapotranspiration. Nat. Clim. Chang., 3, no 1, (2013): 59–62.

Droogers, P., and Allen, R.G. Estimating reference evapotranspiration under inaccurate data conditions. Irrig. Drain. Syst., 16, (2002): 33–45.

Dyck, S. Overview on the present status of the concepts of water balance models. IAHS Publ., 148, (1985): 3–19.

Er-Raki, S., Chehbouni, A., Khabba, S., Simonneaux, V., Jarlan, L., Ouldbba, A., Rodriguez, J.C., and Allen, R. Assessment of reference evapotranspiration methods in semi-arid regions: Can weather forecast data be used as alternate of ground meteorological parameters? J. Arid Environ., 74, (2010): 1587–1596.

Eslamian, S.S., Gohari, S.A., Zareian, M.J., and Firoozfar, A. Estimating Penman—Monteith reference evapotranspiration using artificial neural networks and genetic algorithm: A case study. Arab. J. Sci. Eng., 37, (2012): 935–944.

Evett, S.R., and Lascano, R.J. ENWATBAL.BAS: A mechanistic evapotranspiration model written in compiled basic. Agron. J., 85, (1993): 763–772.

Farah, H.O., Bastiaanssen, W.G.M., and Feddes, R.A. Evaluation of the temporal variability of the evaporative fraction in a tropical watershed. Int. J. Appl. Earth Obs. Geoinf., 5, (2004): 129–140.

Farzanpour, H., Shiri, J., Sadraddini, A.A., and Trajkovic, S. Global comparison of 20 reference evapotranspiration equations in a semi-arid region of Iran. Hydrol. Res., 50, no 1, (2019): 282–300.

Fausset, L.V. Fundamentals of Neural Networks: Architectures, Algorithms and Applications. Prentice Hall, Upper Saddle River, (1994).

Federer, C.A., Vörösmarty, C., and Fekete, B. Intercomparison of methods for calculating potential evaporation in regional and global water balance models. Water Res. Res., 32, (1996): 2315–2321.

Felzer, B.S., Cronin, T.W., Melillo, J.M., Kicklighter, D.W., and Schlosser, C.A. Importance of carbon-nitrogen interactions and ozone on ecosystem hydrology during the 21st century. J. Geophys. Res., 114, no G1, (2009): 1–10.

Fernandes, C., Cora, J.E., and Araujo, J.A.C. Reference evapotranspiration estimation inside greenhouse. Sci. Agric., 60, no 3, (2003): 591–594.

Ferreira, M.I., Silvestre, J., Conceiçao, N., and Malheiro, A.C. Crop and stress coefficients in rain fed and deficit irrigation vineyards using sap flow techniques. Irrig. Sci., 30, (2012): 433–447.

Fontenot, R.L. An evaluation of reference evapotranspiration models in Louisiana. Louisiana State University and Agricultural & Mechanical College, Baton Rouge, LA, USA. (2004).

Frevert, D.K., Hill, R.W., and Braaten, B.C. Estimation of FAO evapotranspiration coefficients. J. Irrig. Drain. ASCE, 109, (IR2), (1983): 265–270.

Fynn, R.P., Al-Shooshan, A., Short, T.H., and McMahon, R.W. Evapotranspiration measurements and modeling for a potted chrysanthemum crop. Trans. the ASAE, 36, no 6, (1993): 1907–1913.

Gao, Y., Long, D., and Li, Z. Estimation of daily evapotranspiration from remotely sensed data under complex terrain over the Upper Chao River Basin in North China. Int. J. Remote Sens., 29, (2008): 3295–3315.

Gardelin, M., and Lindstrom, G. Priestley-Taylor evapotranspiration in HBV-simulations. Nord. Hydrol., 28, no 4/5, (1997): 233–246.

Garson, G.D. Interpreting neural-network connection weights. Exp. Syst. Appl., 6, (1991): 47–51.

Gebhart, S., Radoglou, K., Chalivopoulos, G., and Matzarakis, A. Evaluation of potential evapotranspiration in central Macedonia by EmPEst. In: Helmis, C., and Nastos, P. (eds.). Advances in Meteorology, Climatology and Atmospheric Physics, Vol. 2. Springer Atmospheric Sciencies, Springer, Berlin Heidelberg, (2013): 451–456.

Gebremichael, M., Wang, J., and Sammis, T.W. Dependence of remote sensing evapotranspiration algorithm on spatial resolution. Atmos. Res., 96, (2010): 489–495.

Gedney, N., Cox, P.M., Betts, R.A., Boucher, O., Huntingford, C., and Stott, P.A. Detection of a direct carbon dioxide effect in continental river runoff records. Nature, 439, no 7078, (2006): 835–838.

Glenn, E., Huete, A., Nagler, P., and Nelson, S. Relationship between remotely sensed vegetation indices, canopy attributes and plant physiological processes: What vegetation indices can and cannot tell us about the landscape. Sensors, 8, no 4, (2008): 2136–2160.

Gonzalez-Dugo, M.P., Neale, C.M.U., Mateos, L., Kustas, W.P., and Li, F. Comparison of remote sensing-based energy balance methods for estimating crop evapotranspiration. Rem. Sens. for Agric. Ecosyst. Hidrol.VIII, 6359, (2006): 218–226.

Gonzalez-Dugo, M.P., Neale, C.M.U., Mateos, L., Kustas, W.P., Prueger, J.H., Anderson, M.C., and Li, F. A comparison of operational remote sensing-based models for estimating crop Evapotranspiration. Agric. For. Meteorol., 149, (2009): 1843–1853.

Goodarzi, M., and Eslamian, S. Performance evaluation of linear and nonlinear models for the estimation of reference evapotranspiration. Int. J. Hydrol. Sci. Tech., 8, no 1, (2018): 1–15.

Gowda, P.H., Senay, G.B., Howell, T.A., and Marek, T.H. Lysimetric evaluation of simplified surface energy balance approach in the Texas high plains. Appl. Eng. Agric., 25, (2009): 665–669.

Grismer, M.E., Orang, M., Snyder, R., and Matyac, R. Pan Evaporation to reference evapotranspiration conversion methods. J. Irrig. Drain. Eng., 128, no 3, (2002): 180–184.

Gurney, R.J., and Hsu, A.Y. Relating evaporative fraction to remotely sensed data at FIFE site. In: Symposium on FIFE: Fist ISLSCP Field Experiment, February 7–9. American Meteorological Society, Boston, USA, (1990): 112–116.

Hamon, W.R. Estimating potential evapotranspiration. Trans. Am. Soc. Civil Eng., 128, no 1, (1963): 324–338.

Hansen, S. Estimation of potential and actual evapotranspiration: Paper presented at the nordic hydrological conference (Nyborg, Denmark, August-1984). Hydrol. Res., 15, no 4–5, (1984): 205–212.

Hargreaves, G.H. Moisture availability and crop production. ASCE Trans., 18, no 5, (1975): 980–984.

Hargreaves, G.H. Defining and using reference evapotranspiration. J. Irrig. Drain. Eng., 120, no 6, (1994): 1132–1139.

Hargreaves, G.H., and Allen, R.G. History and evaluation of Hargreaves evapotranspiration equation. J. Irrig. Drain. Eng., 129, no 1, (2003): 53–63.

Hargreaves, G.L., Hargreaves, G.H., and Riley, J.P. Agricultural benefits for Senegal River basin. J. Irrig. Drain. Engr., ASCE, 111, no 2, (1985): 113–124.

Hargreaves, G.H., and Samani, Z.A. Reference crop evapotranspiration from temperature. Appl. Eng. Agric., 1, no 2, (1985): 96–99.

Hess, A., Wadzuk, B., and Welker, A. Predictive evapotranspiration equations in rain gardens. J. Irrig. Drain. Eng., 145, no 7, (2019): 04019010.

Hillel, D. Environmental Soil Physics. Academic Press, London, UK/San Diego, CA, (1998).

Huo, Z., Feng, S., Kang, S., and Dai, X. Artificial neural network models for reference evapotranspiration in an arid area of northwest China. J. Arid Environ., 82, (2012): 81–90.

Idso, S.B., Reginato, R.J., Jackson, R.D., Kimball, B.A., and Nakayama, F.S. Three stages of drying of a field soil. Soil Sci. Soc. Am. J., 38, (1974): 831–837.

Intergovernmental Panel on Climate Change. Climate Change 2007: The Physical Science Basis Contribution of Working Group I to the Fourth Assessment Report of the Intergovernmental Panel on Climate Change. Cambridge Univ. Press, Cambridge, UK, (2007).

Iqbal, M. An Introduction to Solar Radiation. Academic Press, Toronto, Canada, (1983).

Irmak, S., Irmak, A., Allen, R.G., and Jones, J.W. Solar and net radiation based equations to estimate reference evapotranspiration in humid climates. J. Irrig. Drain. Eng., 129, no 5, (2003b): 336–347.

Irmak, S., Irmak, A., Jones, J.W., Howell, T.A., Jacobs, J.M., Allen, R.G., and Hoogenboom, G. Predicting daily net radiation using minimum climatological data. J. Irrig. Drain. Eng., 129, no 4, (2003a): 256–269.

Isaaks, E.H., and Srivastava, R.M. An Introduction to Applied Geostatistics. Oxford University Press, NewYork, USA, (1989).

Jain, S.K., Nayak, P.C., and Sudheer, K.P. Models for estimating evapotranspiration using artificial neural network, and their physical interpretation. Hydrol. Process., 22, no 13, (2008): 2225–2234.

Jensen, D.T., Hargreaves, G.H., Temesgen, B., and Allen, R.G. Computation of ET_0 under non ideal conditions. J. Irrig. Drain. Eng., 123, (1997): 394–400.

Jensen, M.E. Empirical methods of estimating or predicting evapotranspiration using radiation. In: Evapotranspiration and Its Role in Water Resources Management. American Society of Agricultural Engineers, Chicago, USA, (1966): 49–53.

Jensen, M.E. Water consumption by agricultural plants. In: Kozlowski, T.T. (ed.). Water Deficits and Plant Growth, Vol. 2. Academic Press, New York, USA, (1968): 1–22.

Jensen, M.E., Burman, R.D., and Allen, R.G. Evapotranspiration and irrigation water requirements. In: ASCE Manuals and Reports on Engineering Practice. No. 70. (1990).

Jensen, M.E., and Haise, H.R. Estimating evapotranspiration from solar radiation. J. Irrig. Drain., 89, no 4, (1963): 15–41.

Jones, H.G., and Tardieu, F. Modelling water relations of horticultural crops: A review. Sci. Hortic., 74, (1998): 21–46.

Julien, Y., Sobrino, J.A., Mattar, C., Ruescas, A.B., Jiménez-Muñoz, J.C., Sòria, G., Hidalgo, V., Atitar, M., Franch, B., and Cuenca, J. Temporal analysis of Normalized Difference Vegetation Index (NDVI) and Land Surface Temperature (LST) parameters to detect changes in the Iberian Land cover between 1981 and 2001. Int. J. Remote Sens., 32, (2011): 2057–2068.

Kamali, M.E., Nazari, R., Faridhosseini, A., Ansari, H., and Eslamian, S. The determination of reference evapotranspiration for spatial distribution mapping using geostatistics. Water Resour. Manage., 29, (2015): 3929–3940.

Kannan, N., White, S.M., Worrall, F., and Whelan, M.J. Sensitivity analysis and identification of the best evapotranspiration and runoff options for hydrological modeling in SWAT-2000. J. Hydrol., 332, no 3/4, (2007): 456–466.

Kerridge, B.L., Hornbuckle, J.W., Christen, E.W., and Faulkner, R.D. Using soil surface temperature to assess soil evaporation in a drip irrigated vineyard. Agric. Water Manage., 116, (2013): 128–141.

Khaledian, Y., Kiani, F., and Ebrahimi, S. The effect of land use change on soil and water quality in northern Iran. J. Mount. Sci., 9, (2012): 798–816.

Khu, S.T., Liong, S.Y., Babovic, V., Madsen, H., and Muttil, N. Genetic programming and its application in real-time runoff forming. J. Am. Water Res. Assoc., 37, no 2, (2001): 439–451.

Kisi, O. Evapotranspiration modeling from climatic data using a neural computing technique. Hydrol. Process., 21, (2007): 1925–1934.

Kool, D., Agama, N., Lazarovitcha, N., Heitmanc, J.L., Sauerd, T.J., and Ben-Gal, A. A review of approaches for evapotranspiration partitioning. Agr. For. Meteorol., 184, (2014): 56–70.

Koza, J.R. Genetic Programming: On the Programming of Computers by Means of Natural Selection. The MIT Press, Cambridge, MA, USA, (1992).

Krishna, P.A. Evapotranspiration and agriculture—A review. Agric. Rev., 40, (2018): 1–11.

Kurtulus, B., and Razack, M. Modeling daily discharge responses of a large karstic aquifer using soft computing methods: Artificial neural network and neuro-fuzzy. J. Hydrol., 381, (2010): 101–111.

Kustas, W.P., and Agam, N. Soil evaporation. In: Wang, Y.Q. (ed.). Encyclopedia of Natural Resources. Taylor & Francis, New York, USA, (2013).

Kustas, W.P., Hatfield, J., and Prueger, J.H. The Soil Moisture Atmosphere Coupling Experiment (SMACEX): Background, hydrometerological conditions and preliminary findings. J. Hydrometeorol., 6, (2005): 791–804.

Kustas, W.P., and Norman, J.M. A two-source approach for estimating turbulent fluxes using multiple angle thermal infrared observations. Water Resour. Res., 33, (1997): 495–1508.

Kustas, W.P., and Norman, J.M. Evaluation of soil and vegetation heat flux predictions using a simple two-source model with radiometric temperatures for partial canopy cover. Agric. For. Meteorol., 94, (1999): 13–29.

Landeras, G., Ortiz-Barredo, A., and López, J.J. Comparison of artificial neural network models and empirical and semi-empirical equations for daily reference evapotranspiration estimation in the Basque Country (Northern Spain). Agr. Water Manag., 95, no 5, (2008): 553–565.

Lang, D., Zheng, J., Shi, J., Liao, F., Ma, X., Wang, W., Chen, X., and Zhang, M. A comparative study of potential evapotranspiration estimation by eight methods with FAO Penman—Monteith method in southwestern China. Water, 9, no 10, (2017): 734–752.

Lascano, R.J., Van Bavel, C.H.M., Hatfield, J.L., and Upchurch, D.R. Energy and water balance of a sparse crop: Simulated and measured soil and crop evaporation. Soil Sci. Soc. Am. J., 51, (1987): 1113–1121.

Lemon, E.R. The potentialities for decreasing soil moisture evaporation loss. Soil Sci. Soc. Am. J., 20, (1956): 120–125.

Li, F., Kustas, W.P., Prueger, J.H., Neale, C.M.U., and Jackson, J.T. Utility of remote sensing based two-source energy balance model under low and high vegetation cover conditions. J. Hydrometerol., 6, (2005): 878–891.

Li, Y., Qin, Y., and Rong, P. Evolution of potential evapotranspiration and its sensitivity to climate change based on the Thornthwaite, Hargreaves, and Penman—Monteith equation in environmental sensitive areas of China. Atmos. Res., 273, (2022): 106178.

Lingling, Z., Xia, J., Xu, C.-Y., Wang, Z., and Sobkowiak, L. Evapotranspiration estimation methods in hydrological models. J. Geogr. Sci., 23, no 2, (2013): 359–369.

Liou, K.N. An Introduction to Atmospheric Radiation. 2nd ed. Academic Press, San Diego, USA, (2002).

Liu, C.M., and Zhang, D. Temporal and spatial change analysis of the sensitivity of potential evapotranspiration to meteorological influencing factors in China. Acta Geograph. Sin., 66, no 5, (2011): 579–588.

Liu, J., You, Y., Li, J., Sitch, S., Gu, X., Nabel, J.E., Lombardozzi, D., Luo, M., Feng, X., Almut, A., Jain, A.K., Friedlingstein, P., Tian, H., Poulter, B., and Kong, D. Response of global land evapotranspiration to climate change, elevated CO_2, and land use change. Agr. For. Meteorol., 311, (2021): 108663.

López-Urrea, R., de Santa Olalla, F.M., Fabeiro, C., and Moratalla, A. Testing evapotranspiration equations using lysimeter observations in a semiarid climate. Agr. Water Manag., 85, no 1–2, (2006): 15–26.

Lu, J., Sun, G., McNulty, S.G., and Amatya, D.M. A comparison of six potential evapotranspiration methods for regional use in the south-eastern United States. J. Am. Water Res. Assoc., 41, no 3, (2005): 621–633.

Ma, X.N., Zhang, M.J., Wang, S.J., Ma, Q., and Pan, S. Evaporation paradox in the Yellow River Basin. Acta Geograph. Sin., 67, no 5, (2012): 645–656.

Mahringer, W. Verdunstungsstudien am Neusiedler see. Theor. Appl. Climatol., 18, no 1, (1970): 1–20.

Makkink, G.F. Testing the Penman formula by means of lysimeters. J. Inst. Water Eng. Sci., 11, (1957): 277–288.

Makkink, G.F., and Heemst, H.V. Potential evaporation from short grass and water. (1967): 89–96. Mallikarjuna, P., Aruna Jyothy, S., Srinivasa Murthy, D., and Chandrasekhar Reddy,

K. Performance of recalibrated equations for the estimation of daily reference evapo-transpiration. Water Resour. Manage., 28, (2014): 4513–4535.

Mao, J.F., Fu, W.T., Shi, X.Y., Ricciuto, D.M., Fisher, J.B., Dickinson, R.E., Wei, Y.X., Shem, W., Piao, S.L., Wang, K.C., Schwalm, C.R., Tian, H., Mu, M., Arain, A., Ciais, P., Cook, R., Dai, Y., Hayes, D., Hoffman, F.M., Huang, M., Huang, S., Huntzinger, D.N., Ito, A., Jain, A., King, A.W., Lei, H., Lu, C., Michalak, A.M., Parazoo, N., Peng, C., Peng, S., Poulter, B., Schaefer, K., Jafarov, E., Thornton, P.E., Wang, W., Zeng, N., Zeng, Z., Zhao, F., Zhu, Q., and Zhu, Z. Disentangling climatic and anthropogenic controls on global terrestrial evapotranspiration trends. Environ. Res. Lett., 10, no 9, (2015): 1–13.

Mardikis, M.G., Kalivas, D.P., and Kollias, V.J. Comparison of interpolation methods for the prediction of reference evapotranspiration—An application in Greece. Water Resour. Manag., 19, (2005): 251–278.

Martinez-Cob, A. Multivariate geostatistical analysis of evapotranspiration and precipitation in mountain terrain. J. Hydrol., 174, (1996): 19–35.

Matheron, G. Principles of geostatistics. Econ. Geol., 58, (1963): 1246–1266.

Meyer, A. Über einige Zusammenhänge zwischen Klima und Boden in Europa. Chem. Erde, 2, (1926): 209–347.

McMahon, T.A., Peel, M.C., Lowe, L., Srikanthan, R., and Mcvicar, T.R. Estimating actual, potential, reference crop and pan evaporation using standard meteorological data: A pragmatic synthesis. Hydrol. Earth Syst. Sci. Discuss., 9, (2012): 11829–11910.

McNaughton, K.G., and Jarvis, P.G. Predicting effects of vegetation changes on transpiration and evaporation. In: Kozlowski, T.T. (ed.). Water Deficits and Plant Growth, vol. VII. Academic Press, 7, (1983): 1–47.

Mintz, Y., and Walker, G. Global fields of soil moisture and land surface evapotranspiration derived from observed precipitation and surface air temperature. J. Appl. Meteorol., 32, no 8, (1993): 1305–1334.

Moeletsi, M.E., Walker, S., and Hamandawana, H. Comparison of the Hargreaves and Samani equation and the Thornthwaite equation for estimating dekadal evapotranspiration in the free state province, South Africa. Phys. Chem. Earth, Parts A/B/C., 66, (2013): 4–15.

Mohawesh, O.E. Evaluation of evapotranspiration models for estimating daily reference evapotranspiration in arid and semiarid environments. Plant Soil Environ., 57, no 4, (2011): 145–152.

Monteith, J.L. Evaporation and environment. In: Fogg, G.E. (ed.). The State and Movement of Water in Living Organisms. Proc. Symp. Soc. Exp. Biol. Academic Press, (1965): 205–234.

Mpusia, P.T. Comparison of water consumption between greenhouse and outdoor culti-vation. Ph.D. thesis, International Institute for Geo-Information Science and Earth Observation, The Netherland. (2006).

Nalder, I.A., and Wein, R.W. Spatial interpolation of climatic normal: Test of a new method in the Canadian boreal forest. Agric. For. Meteorol., 92, (1998): 211–225.

Neale, C.M.U., Bausch, W.C., and Heermann, D.F. Development of reflectance based crop coefficients for corn. Trans. ASAE, 32, no 6, (1989): 1891–1899.

Neuman, S.P., Feddes, R.A., and Bresler, E. Finite element analysis of two-dimensional flow in soils considering water uptake by roots. I. Theory. Soil Sci. Soc. Am. J., 39, (1975): 224–230.

Nikam, B.R., Kumar, P., Garg, V., Thakur, P.K., and Aggarwal, S.P. Comparative evalua-tion of different potential evapotranspiration estimation approaches. Int. J. Res. Eng. Technol, 3, (2014): 543–552.

Norman, J.M., Anderson, M.C., and Kustas, W.P. Are single-source, remote-sensing surface-flux models too simple? In: D'Urso, G., Osann Fochum, M.A., and Moreno, J. (eds.).

Proceeding of the International Conference on Earth Observation for Vegetation Monitoring and Water Management, Vol. 852. AIP, (2006): 170–177.

Norman, J.M., Kustas, W.P., and Humes, K.S. A two-source approach for estimating soil and vegetation energy fluxes in observations of directional radiometric surface temperature. Agric. For. Meteorol., 77, (1995): 263–293.

Odhiambo, L.O., Yoder, D.C., and Hines, J.W. Optimization of fuzzy evapotranspiration model through neural training with input—Output examples. Trans. ASAE, 44, no 6, (2001): 1625.

Oudin, L., Hervieu, F., Michel, C., Perrin, C., Andréassian, V., Anctil, F., and Loumagne, C. Which potential evapotranspiration input for a lumped rainfall-runoff model? (Part 2): Towards a simple and efficient potential evapotranspiration sensitivity analysis and identification of the best evapotranspiration and runoff options for rainfall-runoff modeling. J. Hydrol., 303, no 1–4, (2005): 290–306.

Oudin, L., Perrin, C., Mathevet, T., Andréassian, V., and Michel, C. Impact of biased and randomly corrupted inputs on the efficiency and the parameters of watershed models. J. Hydrol., 320, no 1–2, (2006): 62–83.

Pan, S., Tian, H., Dangal, S.R., Yang, Q., Yang, J., Lu, C., and Ouyang, Z. Responses of global terrestrial evapotranspiration to climate change and increasing atmospheric CO_2 in the 21st century: Earth's Futur., 3, no 1, (2015): 15–35.

Pandey, P.K., Dabral, P.P., and Pandey, V. Evaluation of reference evapotranspiration methods for the northeastern region of India. J. Soil Water Conserve., 4, no 1, (2016): 56–67.

Parasuraman, K., Elshorbagy, A., and Carey, S.K. Modeling the dynamics of the evapotranspiration using genetic programming. Hydrol. Sci. J., 52, no 3, (2007): 563–578.

Penman, H.L. Natural evaporation from open water, bare soil, and grass. Proceeding of Royal Society of London, UK, A193, (1948): 120–145.

Penman, H.L. Vegetation and hydrology. Soil Sci., 96, no 5, (1963): 357.

Pereira, A.R., Villanova, N., Pereira, A.S., and Baebieri, V.A. A model for the class-A pan coefficient. Agric. Water Manag., 76, (1995): 75–82.

Pereira, L.S., Perrier, A., Allen, R.G., and Alves, I. Evapotranspiration: Concepts and future trends. J. Irrig. Drain. Eng., 125, no 2, (1999): 45–51.

Peterson, T.C., Golubev, V.S., and Groisman, P.Y. Evaporation losing its strength. Nature, 377, (1995): 687–688.

Phillips, D.L., and Marks, D. Spatial uncertainty analysis: Propagation of interpolation errors in spatially distributed models. Ecol. Model., 91, (1996): 213–229.

Piao, S., Friedlingstein, P., Ciais, P., de Noblet-Ducoudre, N., Labat, D., and Zaehle, S. Changes in climate and land use have a larger direct impact than rising CO_2 on global river runoff trends. Proc. Natl. Acad. Sci., 104, no 39, (2007): 15242–15247.

Prentice, I.C., Farquhar, G.D., Fasham, M.J.R., Goulden, M.L., Heimann, M., Jaramillo, V.J., Kheshgi, H.S., Le Quéré, C., Scholes, R.J., and Wallace, D.W.R. The carbon cycle and atmospheric carbon dioxide. In: Climate Change 2001: The Scientific Basis: Contribution of Working Group I to the Third Assessment Report of the Intergovernmental Panel on Climate Change. Cambridge University, Cambridge, United Kingdom and New York, NY, USA, (2001): 183–238.

Priestley, C.H.B., and Taylor, R.J. On the assessment of surface heat flux and evaporation using large scale parameters. Mon. Weather Rev., 100, (1972): 81–92.

Rácz, C., Nagy, J., and Csaba, D.A. Comparison of several methods for calculation of reference evapotranspiration. Acta Silv. Lign. Hung., 9, (2013): 9–24.

Reyes-González, A., Kjaersgaard, J., Trooien, T., Hay, C., and Ahiablame, L. Estimation of crop evapotranspiration using satellite remote sensing-based vegetation index. Adv. Meteorol., (2018): 1–12.

Ritchie, J.T. Model for predicting evaporation from a row crop with incomplete cover. Water Resour. Res., 8, (1972): 1204–1213.

Romanenko, V.A. Computation of the autumn soil moisture using a universal relationship for a large area. Proc. Ukr. Hydrometeorol. Res., Inst., 3, (1961).

Romero, M.G. Daily evapotranspiration estimation by means of evaporative fraction and reference evapotranspiration fraction. Ph.D. Dissertation, Utah State Univ. Logan, Utah, USA. (2004).

Sabziparvar, A.A., Mousavi, R., Marofi, S., Ebrahimipak, N.A., and Heidari, M. An improved estimation of the angstrom—Prescott radiation coefficients for the FAO56 Penman—Monteith evapotranspiration method. Water Resour. Manag., 27, (2013): 2839–2854.

Salas, J.D., Delleur, J.W., Yevjevich, V., and Lane, W.L. Applied Modeling of Hydrologic Time Series. Water Resources Publication, Littleton, CO, USA, (1980).

Samaras, D.A., Rei, A., and Theodoropoulos, K. Evaluation of radiation-based reference evapotranspiration models under different Mediterranean climates in central Greece. Water Res. Manag., 28, (2014): 207–225.

Sarr, M.A., Gachon, P., Seidou, O., Bryant, C.R., Ndione, J.A., and Comby, J. Inconsistent linear trends in Senegalese rainfall indices from 1950 to 2007. Hydrol. Sci. J., 60, (2015): 1538–1549.

Senay, G.B., Bohms, S., Singh, R.K., Gowda, P.H., Velpuri, N.M., Alemu, H., and Verdin, J.P. Operational evapotranspiration mapping using remote sensing and weather datasets: A new parameterization for the SSEB approach. J. Am. Water Resour. Assoc., 49, (2013): 577–591.

Senay, G.B., Budde, M., Verdin, J.P., and Melesse, A. A coupled remote sensing and simplified surface energy balance approach to estimate actual evapotranspiration from irrigated fields. Sensors, 7, (2007): 979–1000.

Setlak, G. The fuzzy-neuro classifier for decision support. Int. J. Inform. Theor. Appl., 15, (2008): 21–26.

Shams, S., Nazemosadat, S., Haghighi, A.K., and Parsa, S.Z. Effect of carbon dioxide concentration and irrigation level on evapotranspiration and yield of red bean. J. Sci. Technol. Greenhouse Cult., 2, no 8, (2012): 1–10.

Shawcroft, R.W., and Gardner, H.R. Direct evaporation from soil under a row crop canopy. Agric. Meteorol., 28, (1983): 229–238.

Shi, X., Mao, J., Thornton, P.E., Hoffman, F.M., and Post, W.M. The impact of climate, CO_2, nitrogen deposition and land use change on simulated contemporary global river flow. Geophys. Res. Lett., 38, no 8, (2011): 1–6.

Shukla, J., and Mintz, Y. Influence of land-surface evapo-transpiration on the Earth's climate. Science, 215, no 4539, (1982): 1498–1501.

Shuttleworth, W.J., Gurney, R.J., Hsu, A.Y., and Ormsby, J.P. FIFE: The variation in energy partition at surface flux sites. IAHS Publ., 186, (1989): 67–74.

Shuttleworth, W.J., and Wallace, J.S. Evaporation from sparse canopy: An energy combination theory. Q. J. Met. Soc., 111, (1985): 839–855.

Šimůnek, J., Šejna, M., Saito, H., Sakai, M., and Van Genuchten, M.T. The HYDRUS-1D software Package for Simulating the Movement of Water, Heat, and Multiple Solutes in Variably Saturated Media, version 4.08. HYDRUS Software Series 3. Department of Environmental Sciences, University of California Riverside, Riverside, CA, USA, (2008).

Song, Z.W., Zhang, H.L., Snyder, R.L., Anderson, F.E., and Chen, F. Distribution and trends in reference evapotranspiration in the North China: Plain. J. Irrig. Drain. Eng., 136, (2010): 240–247.

Stanghellini, C. Transpiration of greenhouse crops: An aid to climate management. Ph.D. thesis, Wageningen Agricultural Univ., Wageningen, The Netherland. (1987).

Stephens, J.C. Discussion of "Estimating evaporation from insolation". J. Hydr. Div., 91, no 5, (1965): 171–182.

Stephens, J.C., and Stewart, E.H. A comparison of procedures for computing evaporation and evapotranspiration. In: General Assembly of Berkeley, Vol. 62. International Association of Hydrological Sciences, Berkeley, USA, (1963): 123–133.

Stull, R.B. An Introduction to Boundary Layer Meteorology. Kluwer Academic Publishers, Boston, USA, (1988).

Sudheer, K.P., Gosain, A.K., and Ramasastri, K.S. Estimating actual evapotranspiration from limited climatic data using neural computing technique. J. Irrig. Drain. Eng. ASCE, 129, (2003): 214–218.

Sumner, D.M., and Jacobs, J.M. Utility of Penman—Monteith, Priestley—Taylor, reference evapotranspiration, and pan evaporation methods to estimate pasture evapotranspiration. J. Hydrol., 308, (2005): 81–104.

Swelam, A., Jomaa, I., Shapland, T., Snyder, R.L., and Moratiel, R. Evapotranspiration response to climate change. In: XXVIII International Horticultural Congress on Science and Horticulture for People (IHC2010), International Symposium on 922. (2010): 91–98.

Szász, G. A potenciális párolgás meghatározásának új módszere. [New method for calculating potential evapotranspiration]. Hidrológiai Közlöny, (1973): 435–442. (in Hungarian).

Tabari, H. Evaluation of reference crop evapotranspiration equations in various climates. Water Res. Manag., 24, (2010): 2311–2337.

Tabari, H., Grismer, M.E., and Trajkovic, S. Comparative analysis of 31 reference evapotranspiration methods under humid conditions. Irrig. Sci., 31, (2012): 107–117.

Tabari, H., Talaee, P.H., Nadoushani, S.M., Willems, P., and Marchetto, A. A survey of temperature and precipitation based aridity indices in Iran. Quater. Int., 345, (2014): 158–166.

Temeepattanapongsa, S., and Thepprasit, C. Comparison and recalibration of equations for estimating reference crop evapotranspiration in Thailand. Agric. Nat. Resour., 49, no 5, (2015): 772–784.

Temesgen, B., Allen, R.G., and Jensen, D.T. Adjusting temperature parameters to reflect wellwater conditions. J. Irrig. Drain. Engrg., 125, no 1, (1999): 26–33.

Thejll, P., and Schmith, T. Limitations on regression analysis due to serially correlated residuals: Application to climate reconstruction from proxies. J. Geophys. Res., 110, no D18, (2005): 103.

Thornthwaite, C.W. The moisture-factor in climate. Eos, Trans. Am. Geophys. Union, 27, no 1, (1946): 41–48.

Thornthwaite, C.W. An approach toward a rational classification of climate. Geogr. Rev., 38, no 1, (1948): 55–94.

Timmermans, W.J., Kustas, W.P., Anderson, M.C., and French, A.N. An intercomparison of the Surface Energy Balance Algorithm for Land (SEBAL) and the Two Sources Energy Balance (TSEB) modeling schemes. Remote Sens. Environ., 108, (2007): 284–369.

Trajkovic, S. Hargreaves versus Penman-Monteith under humid conditions. J Irrig. Drain. Eng., 133, no 1, (2007): 38–42.

Trajkovic, S., and Kolakovic, S. Evaluation of reference evapotranspiration models under humid conditions. Water Resour. Manag., 23, no 14, (2009): 3057–3067.

Trajkovic, S., Todorovic, B., and Stankovic, M. Forecasting reference evapotranspiration by artificial neural networks. J. Irrig. Drain. Eng., 129, no 6, (2003): 454–457.

Troch, P.A., Carrillo, G., Sivapalan, M., Wagener, T., and Sawicz, K. Climate-vegetationsoil interactions and long-term hydrologic partitioning: Signatures of catchment coevolution. Hydrol. Earth Syst. Sci., 17, no 6, (2013): 2209–2217.

Valipour, M., Bateni, S.M., Gholami Sefidkouhi, M.A., Raeini-Sarjaz, M., and Singh, V.P. Complexity of forces driving trend of reference evapotranspiration and signals of climate change. Atmosphere, 11, no 10, (2020): 1081.

Van Halsema, G.E., and Vincent, L. Efficiency and productivity terms for water management: A matter of contextual relativism versus general absolutism. Agric. Water Manage., 108, (2012): 9–15.

Walker, G.K. Measurement of evaporation from soil beneath crop canopies. Can. J. Soil Sci., 63, (1983): 137–141.

Wanniarachchi, S., and Sarukkalige, R. A review on evapotranspiration estimation in agricultural water management: Past, present, and future. Hydrology, 9, no 7, (2022): 123.

Weber, D., and England, E. Evaluation and comparison of spatial interpolators II., Math. Geol., 26, (1994): 589–603.

Weiss, O., Minixhofer, P., Scharf, B., and Pitha, U. Equation for calculating evapotranspiration of technical soils for urban planting. Land, 10, no 6, (2021): 622.

Wen, L. Reconstruction natural flow in a regulated system, the Murrumbidgee River, Australia, using time series analysis. J. Hydrol., 364, (2009): 216–226.

Wilcox, B.P., Breshears, D.D., and Seyfried, M.S. Water balance on rangelands. In: Stewart, B.A., and Howell, T.A. (eds.). Encyclopedia of Water Science. Marcel Dekker, Inc., New York, USA, (2003): 791–794.

Wilson, K.B., Hanson, P.J., Mulholland, P.J., Baldocchi, D.D., and Wullschleger, S.D. A comparison of methods for determining forest evapotranspiration and its components: Sap-flow, soil water budget, eddy covariance and catchment water balance. Agric. For. Meteorol., 106, (2001): 153–168.

WMO measurement and estimation of evaporation and evapotranspiration. In: Technical Paper (CIMO-Rep.) No. 83. Genova, Switzerlands. (1966).

Xu, C.-Y., Gong, L., Jiang, T., Chen, D., and Singh, V.P. Analysis of spatial distribution and temporal trend of reference evapotranspiration and pan evaporation in Changjiang (Yangtze River) catchment. J. Hydrol., 327, (2006): 81–93.

Xu, C.-Y., and Singh, V.P. Cross comparison of empirical equations for calculating potential evapotranspiration with data from Switzerland. Water Resour. Manag., 16, (2002): 197–219.

Xu, C.Y., and Singh, V.P. Evaluation and generalization of temperature-based methods for calculating evaporation. Hydrol. Process., 15, no 2, (2000): 305–319.

Xu, C.Y., Singh, V.P., Chen, Y.D., and Chen, D. Evaporation and evapotranspiration. In: Singh, V.P. (ed.). Hydrology and Hydraulics, 1st ed. Water Resources Publications, Colorado, USA, (2008): 229–276.

Xu, Z.X., and Li, J.Y. Estimating basin evapotranspiration using distributed hydrologic model. J. Hydrol. Eng., 8, no 2, (2003): 74–80.

Yates, D., and Strzepek, K. Potential evapotranspiration methods and their impact on the assessment of river basin runoff under climate change. International Institute of Applied Systems Analysis Working Papers. (1994).

Yin, Y., Wu, S., Zheng, D., and Yang, Q. Radiation calibration of FAO56 Penman—Monteith model to estimate reference crop evapotranspiration in China. Agric. Water. Manag., 95, (2008): 77–84.

Yoder, R.E., Odhiambo, L.O., and Wright, W.C. Evaluation of methods for estimating daily reference crop evapotranspiration at a site in the humid Southeast United States. Appl. Eng. Agric., 21, no 2, (2005): 197–202.

Zeng, Z.Z., Peng, L.Q., and Piao, S.L. Response of terrestrial evapotranspiration to Earth's greening. Curr. Opin. Environ. Sustain., 33, (2018): 9–25.

Zhang, L., Traore, S., Cui, Y., Luo, Y., Zhu, G., Liu, B., Fipps, G., Karthikeyan, R., and Singh, V. Assessment of spatiotemporal variability of reference evapotranspiration and controlling climate factors over decades in China using geospatial techniques. Agric. Water Manag., 213, (2019): 499–511.

Zhao, C., Nan, Z., and Cheng, G. Methods for estimating irrigation needs of spring wheat in the middle Heihe basin, China. Agric. Water Manag., 75, (2005): 54–70.

Zhou, M. Estimates of evapotranspiration and their implication in the Mekong and Yellow River Basins. Evapotranspiration, (2011): 319–358.

Tables List

Figures List

(Continued)

(*Continued*)

Equations List

(Continued)

(Continued)

(Continued)

Equations	Page	
(70)	36	Szász method (1973), part I
(71)	36	Szász method (1973), part II
(72)	37	Makk–FAO24 (1957; Doorenbos and Pruitt, 1977b), part I
(73)	37	Makk–FAO24, part II
(74)	37	Makk–FAO24, part III
(75)	37	Priestley–Taylor model (1972; McNaughton and Jarvis, 1983), part I
(76)	37	Priestley–Taylor, part II
(77)	37	Jensen and Haise (1963) method
(78)	37	$K_T = \dfrac{1}{38 - \dfrac{2Elev}{305} + \dfrac{36.5}{\left(e_{T_{max}}^0 - e_{T_{min}}^0\right)}}$
(79)	38, 69, 76	$T_x = 2.5 + 1.4\left(e_{T_{max}}^0 - e_{T_{min}}^0\right) + \dfrac{Elev}{550}$
(80)	38, 76	WMO-1966
(81)	38	Mahringer model (1970)
(82)	43	Penman (1948)
(83)	43	Makkink modified by Doorenbos and Pruitt (1977a)
(84)	44	root mean square error
(85)	44	mean bias error
(86)	44	$t = \sqrt{\dfrac{(n-1)MBE^2}{RMSE^2 - MBE^2}}$
(87)	44	ET_a calculated using lysimeters
(88)	45	temperature at dew point
(89)	45, 65, 77	$ET_0 = \dfrac{\left(\dfrac{\Delta}{\Delta+\gamma}\right)(R_n - G) + K_w\left[\gamma(\Delta+\gamma)\right](a_w + b_w u_2)(e_s - e_a)}{\lambda}$
(90)	46	modified Hargreaves 2
(91)	46	FAO-24 radiation
(92)	46	FAO-24 pan evaporation, part I
(93)	46	FAO-24 pan evaporation, part II
(94)	47	mean absolute error
(95)	47	relative MAE
(96)	47	calibrated Hargreaves–Samani
(97)	48	calibrated Thornthwaite
(98)	48	$ET_{0Base\_method} = m \times ET_{0method} + c_1$
(99)	50, 52	surface energy balance algorithm for land
(100)	50, 52	$G = \left[\dfrac{LST}{\alpha}\left(0.0038\alpha + 0.0074\alpha^2\right)\left(1 - 0.98NDVI^4\right)\right]R_n$

(Continued)

(Continued)

(Continued)

(Continued)

(Continued)

148

(Continued)

Equations	Page	
(197)	82	Calibrated Priestley–Taylor, semiarid
(198)	82	Modified Priestley–Taylor, humid continental climate
(199)	82	Turc, humid
(200)	82	modified TC
(201)	82	Abtew, warm and humid

(202)   83

$$MAE = N^{-1} \sum_{i=1}^{N} |P_i - O_i|$$

(203)   83 $RMAE = (MAE/\bar{O})100$

(204)   83

$$RMSE = \left[N^{-1} \sum_{i=1}^{N} (P_i - O_i)^2 \right]^{0.5}$$

(205)   86

$$EF = 1 - \frac{\sum_{i=1}^{N} (P_i - O_i)^2}{\sum_{i=1}^{N} (\bar{O} - O_i)^2}$$

(206)   87 $\lambda ET = C_c PM_c + C_s PM_s$

(207)   87

$$PM_c = \frac{s(R_n - G) + \left[\rho c_p (VPD) - sr_a^c \left(R_n^s - G \right) \right] / \left(r_a^a + r_a^c \right)}{s + \gamma \left[1 + \frac{r_s^c}{r_a^a + r_a^c} \right]}$$

(208)   87

$$PM_s = \frac{s(R_n - G) + \left[\rho c_p (VPD) - sr_a^s (R_n - G) \left(R_n^s - G \right) \right] / \left(r_a^a + r_a^s \right)}{s + \gamma \left[1 + \frac{r_s^s}{r_a^a + r_a^s} \right]}$$

| (209) | 87 | latent heat from soil |
| (210) | 87 | latent heat of transpiration (Evett & Lascano, 1993) |

(211)   88

$$\lambda T = \left(\psi_s + \psi_{c\_max} - \psi_c \right) 1000\lambda \left(\frac{LAI}{r_{plant\_hyd}} \right)$$

(212)   88

$$\lambda ET = \frac{\lambda \left(AH_c - AH_a \right)}{r_a}$$

(213)   88

$$\lambda E = \frac{\lambda \left(AH_s - AH_c \right)}{r_{scan}}$$

(214)   88

$$\lambda T = \frac{\lambda \left(AH_l - AH_c \right)}{r_{st} + \eta_{bl}}$$

(215)   88 $\lambda E = \lambda E + \lambda T$

(216)   88 $\lambda T = \alpha_{PT} f_g \frac{s}{s+\gamma} R_n^c$

(217)   89

$$K_{cb} = K_{cb(table)} + [0.004(u_2-2) - 0.004(RH_{min} - 45)] \left(\frac{h_p}{3} \right)^{0.3}$$

(218)   89 $K_e = K_r(K_{c\_max} - K_{cb}) \le f_{ew}K_{c\_max}$

(Continued)

(Continued)

Equations	Page	
(219)	89	$E = -K\dfrac{\partial h}{\partial x} - K \le E_{pot}$ x = L
(220)	89	surface boundary pressure head
(221)	90	$T = \displaystyle\int_{L_R} S(h,h_\varphi,x)dx = T_{pot}\int_{L_R} \alpha(h,h_\varphi,x)b(x)dx$
(222)	92, 105	$ET_0 = (SWC_{to} - SWC_{tl}) + I - D$
(223)	92	Stanghellini
(224)	92	$ET_o = \dfrac{1}{\lambda}\dfrac{\Delta(R_n - G) + K_t\dfrac{VPD\rho C_p}{r_a}}{\Delta + \left(1 + \dfrac{r_c}{r_a}\right)}$
(225)	93	Fynn
(226)	93	$EF = \dfrac{\displaystyle\sum_{i=1}^{n}(O_i - \bar{O})^2 - \sum_{i-1}^{n}(O_i - E_i)^2}{\displaystyle\sum_{i=1}^{n}(O_i - \bar{O})^2}$
(227)	94	$xn_{i,k} = \dfrac{x_{i,k} - m_k}{SD_k}$
(228)	95	$R^2 = \dfrac{E_o - E}{E_o}$
(229)	95	$\delta = \sqrt{\displaystyle\sum_{i=1}^{n}\left(x_i - x_{ij}\right)}$
(230)	96	$f(\delta_i) = Exp(-\lambda\delta_i^2)$
(231)	96	$z_k = \displaystyle\sum_{j=1}^{j} b_{jk} y_j$
(232)	96	$RMSE = \sqrt{\dfrac{1}{N}\displaystyle\sum_{i=1}^{N}(y_t - y_o)^2}$
(233)	96	$R^2 = 1 - \dfrac{\Sigma y_t - y_o}{\Sigma y_t^2 - \dfrac{\Sigma y_o^2}{n}}$
(234)	100	multiple linear regression format
(235)	100	genetic programming model MLR
(236)	102	$ET_0 = a + bE_{model}$
(237)	102	$MBE = \dfrac{\displaystyle\sum_{i=1}^{n}\left(E_{Modeli} - E_{PM56i}\right)}{n}$

(Continued)

(*Continued*)

Equations	Page	
(238)	102	$$\text{MAE} = \left| \frac{\sum_{i=1}^{n}\left(E_{modeli} - E_{PM56i}\right)}{n} \right|$$
(239)	102	$$\text{RMSE} = \sqrt{\frac{\sum_{i=1}^{n}\left(E_{modeli} - E_{PM56i}\right)^2}{n}}$$
(240)	106	fuzzy linear regression
(241)	106	$\tilde{Y} = (p_0,c_0) + (p_1,c_1)x_{i1} + \ldots + (p_n,c_n)x_{in}$
(242)	107	ETo = (0.5431, 0) + (0.1733, 0.1271) Tmin + (0.08652, 0) Tmax + (0, 0.03544) RHmean + (0.0588, 0) U + (0.165, 0) R
(243)	107	ETo = (0.760) + (0.1833, 0.1684) Tmin + (0.089, 0) Tmax + (0, 0.1653) RHmin + (0,0) RHmax + (0.1975, 0) u + (0.1938, 0.0847) R
(244)	107	ETo = (0.0734, 0) + (0.1637, 0.1356) Tmean + (0.00689, 0.239) RHmin + (0, 0) RHmax + (0.3668, 0.093) u + (0.1813, 0) R
(245)	107	$$\text{RMSE} = \sqrt{\frac{1}{N}\sum_{i=1}^{n}(X_k - Y_k)^2}$$
(246)	107	$$R^2 = \left[\frac{\sum_{K=1}^{n}(X_k - X)(Y_k - Y)}{\sum_{K=1}^{n}(X_k - X)^2 \sum_{Y=1}^{N}(Y - Y)^2} \right]^2$$
(247)	111	$$y = \frac{ax^2 + bx + c}{d}$$
(248)	111	E1 (unplanted technical soil A; Weiss et al. (2021)
(249)	111	E2 (planted soil A with Sedum floriferum)
(250)	111	E3 (planted soil A with Geranium x cantabrigiense)
(251)	112	E4 (unplanted technical soil B)
(252)	112	E5 (planted soil A with Sedum floriferum)
(253)	112	E6 (planted soil B with Geranium x cantabrigiense)
(254)	112	E7 (unplanted technical soil C)
(255)	112	E8 (unplanted technical soil D)
(256)	112	E9 (unplanted technical soil E)
(257)	113	E10 (unplanted technical soil F)
(258)	116	ET calculated using FAO56PM for canopy conductance and resistance, part I
(259)	116	FAO56PM ET, part II
(260)	116	FAO56PM ET, part III
(261)	116	$g_s = \max(g_{max}r_{corr}bf\,(ppdf)\,f(T_{min})\,f(vpd)\,f(CO_2),\,g_{min})$
(262)	116	$$r_{corr} = \left(\frac{T_{day} + 273.15}{293.15}\right)^{1.75}\left(\frac{101300}{p}\right)$$

(*Continued*)

(Continued)

(263)   116 $f(ppdf) = \dfrac{ppdf}{75 + ppdf}$

(264)   116 $b_i = 1 \longrightarrow \psi_i > \psi_{open}$

(265)   116 $\dfrac{\psi_{open} - \psi}{\psi_{open} - \psi_{close}} \psi_{close} \leq \psi_i \leq \psi_{open}$

(266)   116 $0 \longrightarrow \psi_i < \psi_{close}$

(267)   116

$$B = \sum_{j=1}^{10}\left(root_i \,\frac{\theta_{sat,j} - \theta_{ice,j}}{\theta_{sat,j}}\, b_i \right)$$

$f(T_{min}) = 1 \; Tmin > 0\,°C$

(268)   117 $1 + 0.125 T_{min} - 8\,°C \leq T_{min} \leq 0°$

$0 \; T_{min} < -8\,°C$

$f(vpd) = 1 \; vpd < vpd_{open}$

(269)   117 $\dfrac{vpd_{close} - vpd}{vpd_{close} - vpd_{open}} \; vpd_{open} \leq vpd \leq vpd_{close}$

(270)   117 $f(CO_2) = -0.001 CO_2 + 1.35$

(271)   117 $EVAP = pet_{PM} e^{-0.6 LAI}$

(272)   117 $ET_{CO2} = \dfrac{ET_{clm + CO_2} - ET_{clm}}{CO_2\,concentration\,(ppm)}$

Index

Printed in the United States
by Baker & Taylor Publisher Services